Climate Change and the Nation State

ANATOL LIEVEN

Climate Change and the Nation State

The Realist Case

ALLEN LANE
an imprint of
PENGUIN BOOKS

ALLEN LANE

UK | USA | Canada | Ireland | Australia
India | New Zealand | South Africa

Allen Lane is part of the Penguin Random House group of companies
whose addresses can be found at global.penguinrandomhouse.com

First published in the USA by Oxford University Press 2020
First published in Great Britain by Allen Lane 2020
001

Printed and bound in Great Britain by Clays Ltd, Elcograf S.p.A.

A CIP catalogue record for this book is available from the British Library

ISBN: 978-0-241-39407-6

For Katya, who will have to live with this mess.

But one of the first and most leading principles on which the commonwealth and the laws are consecrated, is lest the temporary possessors and life-renters in it, unmindful of what they have received from their ancestors, or of what is due to their posterity, should act as if they were the entire masters; that they should not think it amongst their rights to cut off the entail, or commit waste on the inheritance, by destroying at their pleasure the whole original fabric of their society.

—Edmund Burke, *Reflections on the Revolution in France,* edited by Frank M. Turner (Yale University Press, New Haven CT 2003), page 81

CONTENTS

ACKNOWLEDGMENTS

I should like to thank Simon Winder at Penguin and David McBride at Oxford University Press for their interest in my work over the years and their support for what initially seemed an unpopular and unfashionable project on climate change. Holly Mitchell and Jeremy Toynbee gave invaluable help with the editing process. I am also grateful to Natasha Fairweather, Matthew Marland, and their colleagues at Rogers, Coleridge and White for all their help. Katharine Danilowicz gave valuable assistance with the research. Michael Klare, Matthew Dal Santo, Ashraf Hayat, Andrew Holland, Michael Greco, Geoff Kemp, and Joshua Mitchell gave most kind and helpful advice on the book, but are not of course responsible for its contents.

Last I must thank my wife, Sasha, and our children, Misha and Katya. They are largely responsible for this book, since if they were not around, I would very probably have lapsed into paralyzed despair on the subject of climate change and the future of Western democracy; but no decent parent has that option.

INTRODUCTION

We are bound together in a single garment of destiny.

—Martin Luther King Jr.[1]

This book has its origins in a growing sense of alarm, of frustration, and of futility. As international efforts to reduce emissions have failed repeatedly to meet their targets, even as warnings by experts about the existential dangers of climate change and the need for haste have grown, I developed a stronger and stronger sense of the comparative irrelevance of most of the issues on which I have been working in the areas of international relations and security studies.

A revelatory moment came when I was researching the growing tension between the United States and China over the Chinese military occupation of reefs and sandbanks in the South China Sea. I suddenly realized that as a long-term issue these places will be meaningless for both sides: because if nations, and China and the United States above all, fail to take action to limit climate change, by the end of this century rising sea levels and intensified typhoons will have put the sources of these tensions under water again.

The rush of Western security establishments toward a "new cold war" with China and Russia (and new US threats of war with Iran) provided an additional impetus to write this book; for in all the innumerable articles and essays on this subject, hardly one has mentioned the destructive effects of hostility between China and the West on international co-operation against climate change.[2]

In my approach to international relations I have always considered myself a realist, though one strongly influenced by the Christian Realist philosophy of Reinhold Niebuhr and the ethical dimensions of the works of Hans Morgenthau and George Kennan. As such, I have never believed in national power as an end in itself but only as a path to the achievement of greater goals, including universal human ones: defense of the Western democracies against serious threats to their security, but also international peace and cooperation through a recognition of the legitimate interests and the power of other states.[3]

This book is therefore in part an appeal to my fellow realists, and to sensible and patriotic policymakers all over the world, to reassess the interests of their nations, in view of the overwhelming evidence for the severe danger posed to the future of those nations by anthropogenic climate change.[4] It is true that climate change is a new issue for realists—but then, it is a new issue for everybody, historically speaking.

Realists have always been chiefly concerned with state interests, and as this book will argue, the long-term interests of the world's great powers are far more threatened by climate change than they are by each other. Realists claim to be concerned with reality, not with vague dreams of a perfect future. They ground their view of reality on a sober study of the evidence: on what is, not what ought to be. Moreover, of the cardinal virtues, the most important for realists are prudence and courage—and if there was ever an issue that demands prudence in judgment and courage in action, it is climate change.[5]

If intelligent realists do not wake up to the threat of climate change today, they will have no choice but to do so tomorrow, when it may be too late to prevent catastrophic results for their nations. This book is mainly directed at audiences in the Western democracies; but this part of it should appeal equally to patriotic elites in Beijing, Moscow, and Delhi. In democracies and especially the United States, strong public recognition by the military of the threat of climate change to the United States is of great importance because of the influence the military has over conservative audiences who are no longer willing to listen to civilian experts. As Nicholas Stern, the leading economist responsible for the British government's report on the economic consequences of climate change, has pointed out, "Messengers matter. Different audiences trust different types of messenger. . . . [C]limate change communication, to be effective, must utilise rhetoric and frames that resonate with the values and emotions that could inspire action."[6]

Some militaries—including intermittently that of the United States—have recognized the threat of climate change; but as Chapter 1 will argue, to date they have paid it only a fraction of the attention that they have devoted to far lesser threats from China, Russia, and even—absurdly—Iran.

We are in fact suffering from a severe case of what has been called "residual elites": ruling groups that have been shaped in particular historical circumstances, to meet particular sets of challenges and opportunities, and who find it difficult or impossible to adapt to new circumstances and to change so as to meet new challenges and opportunities. The more successful and long-lasting a particular ruling order has been, the greater the difficulty it is likely to have in changing itself.

This helps explain the enthusiasm with which Western security elites have embraced the idea of a "new cold war" with Russia and China—an analogy that is both largely false and wholly unnecessary. The tragicomic thing about the exaggeration of the Russian menace in particular is that while presented in terms of threats to Western security, the alleged threat from Russia in fact represents security and comfort for Western establishments: the security of the Cold War models by which they were shaped more than two generations ago.

This saves them from thinking about much greater and much more difficult and painful challenges, which undermine several basic Western assumptions and shibboleths. These challenges to establishment thinking in the West include the need for severe constraints on capitalism and the need to impose strict limits on immigration. By contrast, reading the endless debates about Europe's need to do more for its own security against the threat from Moscow, one can well imagine NATO bureaucrats simply pulling out identical memos from the 1950s, changing a few dates and names, and issuing them again with no further effort.

The Requirements of Action on Climate Change

However, as so often happens (or at least, happens to me), this book grew and developed as it was being written. From an argument about climate change as a security threat, it grew to include arguments about the political cultures and economic systems that will be necessary if Western democracies are to succeed both in limiting climate change and in withstanding its effects.

It also became increasingly clear that the fundamental obstacle to action in developed Western countries is neither technological nor a lack of financial resources. These constraints are of course very serious ones, but as the response to the financial crisis of 2008 showed, in a real crisis, enormous sums of money can be found. Rather, the essential problem is the lack of motivation and mobilization of elites all over the world, and of voters in the West. Efforts to generate this on a sufficient scale have so far manifestly failed.

My study of Pakistan, dating back more than three decades, was a particular spur to my thinking about climate change. On the one hand, Pakistan is one of the major countries of the world most acutely threatened. Dependence on the Himalayan glaciers, on artificial irrigation, and on the monsoon means that dangers which elsewhere are prospective are in Pakistan already imminent. On the other hand, Pakistan is also a case study in how a dysfunctional political system with no sense of common national purpose can find itself incapable of taking action even in the face of such urgent dangers.

The adoption during the writing of this book of the Green New Deal idea by Democratic candidates for the US presidency is a very hopeful sign; but to implement a program of this scale and radicalism, the Democrats will need to win repeated elections by sweeping majorities and create a new dispensation in national politics akin to the original New Deal. This book makes an argument for how such mass public support can be achieved.

Trying to generate sufficient support for serious action against climate change, however, involves pushing against immense gravitational forces: not just resistance to the radical economic reforms that will be required, and the power of vested interests, but against contemporary culture itself, with its intense materialist individualism and shorter and shorter attention spans. It wouldn't be so bad if—as some on the left like to believe—the fight were only against the banks and the oil companies. Twitter, Chirrup, Burble, the Selfie, the Kardashians and their ilk, and the colossus of modern advertising are all in their different ways in the ranks of the enemy. To prevail against this combination, very powerful and united movements will be required.

Appeals to the common interest of humanity and our duty to the planet, though morally valid and a source of inspiration for activists, have simply not been able to generate enough of a mass appeal in any important country to spur action; and to make the pursuit of an ideal world order an obstacle to the achievement of real action against climate change would in fact be to betray the interests of humanity:

> The equation of political moralizing with morality and of political realism with immorality is itself untenable. The choice is not between moral principles and the national interest, devoid of moral dignity, but between one set of principles divorced from political reality and another set of principles derived from political reality.[7]

This recognition led me naturally to a consideration of nationalism as the most powerful source of collective effort in modern history, at least now that religion (at least in the advanced industrialized countries) and communism are (temporarily?) in abeyance. Among the leaders of the great greenhouse gas emitting nations outside the West, Xi Jinping, Narendra Modi, and Vladimir Putin are ruthless but sincere nationalists, dedicated to the power and survival of their nations, which are central to their own identities and the interests of their ruling oligarchies. Convince them that something threatens those nations, and they will act—as Chinese policy on climate change already demonstrates.

At present, mass nationalism has indeed been chiefly used as a weapon against action on climate change—as the melancholy examples of Donald Trump in the United States and Jair Bolsonaro in Brazil demonstrate. Bolsonaro has made particular play with a latter-day "Third World" discourse of nationalist resentment at being lectured to by wealthy Western countries on the preservation of the Amazon. Nonetheless, there is no innate and inevitable link between nationalism and opposition to action on climate change. As Israeli policies on water conservation and Chinese and Indian policies on alternative energy both in their different ways show, nationalism and considerations of national security can also be mobilized behind effective conservation policies.

The choice then is between stupid, short-sighted versions of nationalism and intelligent, far-sighted ones. After all, while the destruction of the Amazon rainforest would be a catastrophe for humanity in general, it would also be first and foremost a catastrophe for Brazil. The Amazon is also a great Brazilian national symbol, the destruction of which would culturally maim the nation. To defeat the Trumps and Bolsonaros on climate change, it is necessary to seize the weapon of nationalism from their hands and turn it against them.

I contemplated using the word "patriotism" rather than "nationalism" in this book, as a less controversial term.[8] However, this is in the end a meaningless distinction, and a somewhat pusillanimous equivocation. If one speaks of Indian civic nationalism, there is no reason not to do so for the United States, France, or Sweden. Moreover, the leading thinkers who have helped to inspire this book, including Tom Nairn, Yael Tamir, David Miller, Will Kymlicka, and Yascha Mounk, have all written about positive sides of nationalism (not patriotism), and I am happy to follow their example.[9]

Of course, one should be fully and constantly aware of the dreadful forms that nationalism can assume and be careful to guard against them. But then, that is true of every human ideology. Religion can take the form of the Inquisition or the Islamic State. Socialism can become Stalinism or Maoism. Liberalism can become a cover for elitist egotism, exploitation, and kleptocracy. Conservatism can become a cover for stupidity and willful ignorance. No reasonably objective person, however, would say that these possible features in themselves invalidate the good features of these ideologies, or their capacity to learn from each other for the common good: "Any collectivity can provide the institutional structures and the patterns of agency necessary for working out a version of the good life. And any collectivity can display the egoism, arrogance, and general nastiness that we associate today with the rogue nation."[10]

As a journalist in the Caucasus in the 1990s, I witnessed some of the dreadful sides of ethnic nationalism and its capacity to cause conflicts and atrocities. As a journalist and researcher in Pakistan and Afghanistan, however, I have also witnessed how the absence of a strong state nationalism cripples the ability of a state to pursue successful development—and in the worst case can destroy a state altogether. There are no prosperous societies in weak or failed states.[11] This perception has been strengthened still further by recent years spent in the Middle East, watching (this time from a safe distance) the collapse of Syria, Libya, and Yemen.

Most important of all when it comes to the struggle to limit climate change, nationalism is perhaps the only force (other than direct personal concern for children and grandchildren) that can overcome one of the greatest obstacles to serious action; namely, that it requires sacrifices by present generations on behalf of future generations.[12]

The inability of many contemporary economists and philosophers to think in terms of nations, and their attempts to argue about climate change in terms of individual rights, has led them into two intellectual and moral dead ends. The Rational Choice theorists and their like prioritize the rights of existing individuals (or even existing prosperous middle-aged individuals) in ways that are disastrous for our descendants, the survival of our societies, and even potentially our species, and which any previous human culture would have seen as frankly evil. On the other hand, those who do care about the future often find themselves going round and round in circles as they attempt the impossible task of attaching mathematical weight to the relative interests of individuals today, as opposed to those in 50 years' time, 100 years' time, and so on.[13]

To center the argument instead on the interests of nations provides a way out of this impasse. For in the words of Milan Kundera, "A man knows that he is mortal, but he takes it for granted that his nation possesses a kind of eternal life."[14] The central purpose of nationalism is to prolong that life as far as possible into the future. Sacrifices to ensure the future survival of the nation are legitimized and indeed demanded by the fact that previous generations have also sacrificed themselves for this purpose:

> Why do we not exhaust the heritage of the ages, spiritual and material for our immediate pleasure, and let posterity go hang? So far

> as simple rationality is concerned, self-interest can advance no argument against the appetite of present possessors. Yet within some of us, a voice that is not the demand of self-interest or pure rationality says that we have no right to give ourselves enjoyment at the expense of our ancestors' memory and our descendants' prospects. We hold our present advantages only in trust.[15]

For the sacrifices required in the struggle to prevent future climatic disaster and social collapse are real. It is natural and correct that in response to the grotesque exaggerations of right-wing politicians and their corporate funders about the threatened destruction of the economy by a move to alternative energy and energy conservation, environmentalists should stress the great economic benefits that are also involved. The Green New Deal idea rightly places "Green growth" at its core.[16]

It would be a mistake, however, for progressives to deceive the public and themselves that there will be no need for sacrifice. Blaming corporations for their greed and their political manipulation is necessary, but it is also insufficient.[17] Opinion polls show the number of Americans who worry about climate change has increased, and according to one poll reached 69 percent in 2018. An Associated Press poll, however, also indicated that only 28 percent were prepared to pay even an extra $10 a month to combat climate change (while, for example, taken as an average, every US citizen pays more than $200 in taxes per month to fund the US military against largely imaginary threats to the country).[18]

Environmentalists and progressives in general often build their hopes for the future on the belief that the younger generation is more progressive than their elders. This may be true when it comes to recognizing that climate change is a problem, but when it comes to making sacrifices to do something about it, the evidence is ambiguous at best. Thus, in a British survey conducted in 2017, the over-35s were much more likely to support the obligation to pay taxes, while those in the 18–34 age group were more likely to support people keeping what they earned.[19] This can justly be blamed on the universal pressure of capitalist materialism, but the contemporary left's emphasis on individual liberation, identity, selfhood, and victimhood rather than duty, and on the benefits of immigration and diversity rather than community, are also not assets when it comes to asking people to make sacrifices for a common cause.

The need for sacrifice can be seen in almost any issue involving the reduction of CO_2 emissions—and it is this that opponents of taking action have played on above all when it comes to mobilizing voters. In England, regulations requiring higher energy conservation standards for new housing will increase the price of that housing at a time when immigration and uncontrolled property speculation are already pricing a majority of young people out of the housing market in the south of the country.[20]

In France, while one can doubt the purposes and criticize the politics of President Emmanuel Macron's increased tax on diesel fuel that sparked the mass "Yellow Vest" protest movement against him, no one can doubt that this is exactly the kind of tax that will be required if motorists and the car industry are to be shifted away from fossil fuels. In Australia, a population more heavily dependent even than the French on cars will have to pay considerably more for petrol as a result of a carbon tax.

In the developed countries of the West, the rate of carbon gas emissions relative to population and GDP has been falling for several decades, albeit not nearly fast enough. However, this reduction is due only in part to deliberate action to limit emissions. To a very great extent it is due to deindustrialization. And where have the industries gone? To Asia. And where are the goods they produce going? Still very largely to Western consumers, who are thereby simply fueling the carbon emissions of Asia.

When the Democrats come to power in the United States and try to impose much stricter emissions standards on energy use by US industry, this demand is bound sooner or later to be accompanied by rigorous insistence that China and other countries genuinely meet the same standards for products exported to the United States. The EU will do the same (Elizabeth Warren has suggested this as part of her Green New Deal strategy). Politically speaking, this will also help the Democrats address the worries about trade of the US working classes, and win them back from Donald Trump and his like.

It will also require Americans to make sacrifices by paying considerably higher prices for their consumer goods. Such sacrifice by ordinary people can only be demanded as part of a system of social and national solidarity in which the rich have to pay their fair share, and in which the state guarantees the basic living standards of the population as a whole through social welfare and universal health care.

This book argues that a strong civic (as opposed to ethnic) nationalism is also necessary because of the importance of national unity and solidarity in the face of the effects of climate change. While researching this book I came to the gloomy conclusion that because of the failure to act over the past three decades, some severe effects are now inevitable and have indeed already begun. The challenge is to prevent them from becoming catastrophic.

There is of course a danger, a dilemma, and a challenge in appealing to nationalism, well set out by Paul Collier:

> The brute fact is that the domain of public policy is inevitably *spatial*. The political processes that authorise public policy are spatial: national and local elections generate representatives with authority over a territory. . . . The non-spatial political unit is a fantasy, so the only real option is to revive spatial bonds. Unfortunately, given that the most practical unit for most polities is national, we need a sense of shared national identity. But we know that national identities can be toxic. Is it possible to forge bonds that are sufficient for a viable polity yet not dangerous? This is the central question that has to be addressed in social science. On its answer rests the future of our societies.[21]

I entirely agree with this, but I have to add that if states are both to demand the sacrifices necessary to combat climate change and to survive the effects of climate change, it will not be enough for them to be "viable." They, and the national identities that underpin them, will have to be strong. There is no contradiction between this and international cooperation against climate change. To cooperate effectively, nation states need to bring effective powers to the table.[22] In David Miller's apt formulation, "Nations are communities that *do* things together" (my italics).[23]

The social and political danger to Western states is greater in the next decades even than most climate change scientists realize, because the effects of climate change will combine with two other critical challenges for Western societies: automation and artificial intelligence, which threaten the whole contemporary structure of employment, and migration. In combination with white nationalism, mass migration threatens irredeemably to divide societies and paralyze their political systems. Part of the background noise to the writing of this book strongly increased my fears in this regard: not just the Trump administration in the United States and the rise of chauvinist parties in Europe, but the amazing magic show called Brexit, in which a political order once renowned for its pragmatism and common sense transformed itself into play dough before our very eyes.

Populations have become divided in their fundamental understandings of their own national identities: in the United States, believers in a multi-cultural country defined by ideology and defined by multiple identities against believers in a cultural community chiefly defined by a confused appeal to an Anglo-American heritage; in Britain, believers in a multi-cultural, multi-ethnic Britain as part of the European Union against believers in an independent England defined by its own national history. As the miserable examples of Turkey and Egypt demonstrate, it is impossible to make democracy work when at each election, not policies but the very definition of the nation itself is at stake.

Such fractured political systems will have even less ability to do anything serious about anthropogenic climate change. Unless Western democracies can summon up the will to address these challenges, they will ultimately face a choice between authoritarian rule and complete political and social collapse. Having worked in Russia during the near collapse of the state and society in the 1990s, such a scenario is for me not a futurist fantasy but a vivid memory.

We therefore need to develop "sustainable states" in a double and conjoined sense: states that will develop sustainable economies not dependent on fossil fuels and other non-renewable resources; and resilient states that will be able to sustain themselves in the face of the appalling shocks that the next century has in store for us.[24]

Legitimate states exist to defend the interests and security of their citizens and are essential to this task.[25] Yes, states can *sometimes* be monstrous threats to liberty; but the absence of responsible state power is a *permanent* nightmare for anyone without bodyguards or an armed clan to defend themselves and their families: "The state—partial, flawed and often oppressive as it is—is all that stands between us and the unmediated power of money and weapons."[26]

International activist movements and international agreements are certainly vital in the fight against anthropogenic climate change. I am grateful to Greta Thunberg and the School Strike campaign and to Extinction Rebellion, for helping to push governments into paying attention. However, neither such movements, nor international NGOs, nor the "international community" can do anything by themselves. Their goal is to persuade and pressure states to act, and for that there have to be states capable of action. In the Western democracies, this also means governments and political parties capable of winning elections and persuading voters to support climate change action in plebiscites—something that most have so far generally been very unwilling to do when it appears that this will require sacrifices on their part. Indeed, most progressive reforms, including social welfare and women's rights, can only be implemented by effective and legitimate national states.

The environmentalist slogan "Think globally, act locally" actually means, "Think globally, act nationally." Or to adapt David Goodhart's famous formulation: to achieve goals that are in the interests of Everywhere and ardently desired by the Anywheres, you have to rely on Somewhere, and therefore to mobilize support among the Somewheres.[27] The attachment of many Somewheres to their local environment and beloved landscapes is a very good starting point for this.

Talk of the need for nation states to disappear and be replaced by international governance is utterly pointless. It isn't going to happen.[28] If action against climate change depends on the abolition of nation states then there will be no action. Existing nation states may well eventually collapse due to climate change, but the result will be not world government but universal chaos.

The nationalism that underpins strong nation states is also not going to disappear. Predictions that globalization would weaken nationalism have proved almost the exact opposite of the truth. As in the previous great era of modern capitalist globalization between 1871 and 1914, rapid and uncontrolled economic, social, and cultural change is strengthening nationalism, as people look to national identity to preserve some element of inherited culture, and to nation states to give them some protection against capitalist exploitation and uncontrolled movements of transnational finance.

My sense of the importance of strong states backed by strong nationalisms stems in part from my experience of Pakistan. That state—as I predicted in my book of 2011—has proved much stronger than most observers predicted when it comes to surviving and defeating attempts to overthrow it. The problem is that the Pakistani state does not seem capable of doing much more than surviving. For more than 30 years I have seen one vital promised reform after another founder on the reefs of predatory and uneducated political elites; an ignorant and indifferent population; a demoralized, corrupt, and amateurish bureaucracy; a fundamental lack of state legitimacy; and ethnic, social, regional, and sectarian divisions that feed political paralysis and make any collective effort extremely difficult. The dire picture Robert Reich has painted of the consequences of the absence of "a shared sense of responsibility to the common good" are Pakistani reality.[29]

I always said that a more accurate title of my book on Pakistan, *Pakistan: A Hard Country*, would have been *Pakistan Trundles Along*; and the track along which it is trundling is taking it to collapse once the effects of climate change really kick in.[30] We in the West shouldn't feel smug or superior, however. If we fail to limit climate change, Pakistan will only go off the rails a few decades before the West does. Indeed, the collapse of states like Pakistan will in turn help push *us* off the rails.

A perception of the need to strengthen states internally led me to the idea that we need to combine measures to limit climate change with measures to strengthen social solidarity, support employment, and limit tax avoidance by the rich. This necessity forms part of a tradition going back more than 150 years of movements from above and below acting to place limits on capitalism and thereby to save capitalism from itself.

Once again, however, such social programs can only be implemented on a national basis and by strong national states.[31] Ideas of international solidarity involving the transfer of immense resources from developed to poor countries are the purest fantasy (unless they can be convincingly linked to jobs, through massive exports of alternative energy technology).[32] Even within the European Union, large-scale social solidarity across national borders notoriously failed in the wake of the 2008 financial crash. Measures to strengthen *national* social solidarity by contrast are possible, are necessary to rebuild national unity, and are also essential if populations are to be persuaded to make sacrifices to limit climate change.

I therefore strongly support the general thrust of the Green New Deal and analogous proposals in Europe toward the radical reform of capitalism and the expansion of social security. However, as Green thinking grows in the US Democratic Party and Green parties in Germany and elsewhere in Europe rise to replace the Social Democrats as the dominant party on the left, it is ever more important that they should really prioritize the linked goals of action against climate change, social solidarity, and national unity. They should not allow these goals to be harmed by other agendas that sometimes directly contradict them.

My position is close to that of David Rosenberg (writing in the Israeli newspaper *Haaretz*):

> Climate change hits all the [Republicans'] red buttons—massive state intervention and global cooperation led by pointy-headed bureaucrats. . . . Faced with becoming ideological Luddites, they don't just reject the solution to climate change, they reject the science. . . . You shouldn't smugly assume that the left's interest in climate change is entirely grounded in science either. It pushes all their ideological joy buttons. The difference is that recognition of climate change in the end has real science behind it, while rejection is basically the stuff of cranks.[33]

Both sides of the political spectrum suffer from their own versions of ideological conformity and complacency as far as climate change is concerned. Among many on the right, there is a dogmatic attachment to rigid free market economics of a kind that makes serious recognition of the threat of climate change extremely difficult—because there is no way to limit climate change without massive state intervention in the economy. This pseudo-conservative denial of climate change has little to do with the old pragmatic and patriotic traditions of Anglo-American conservatism and reminds me somewhat of the diehard communist officials I interviewed in the last years of the USSR, completely unable to adapt to a changing world.

On the left, too many environmentalists, on the other hand, manage strangely to combine belief in the apocalyptic danger of climate change (often exaggerated as far as the short term is concerned) with a belief that this can be not just checked but transformed in a positive direction by progressive effort undertaken by politically correct and above all (in their own estimation) morally *good* people. This is a vision of a nice, ideologically positive apocalypse inhabited by diverse but mutually respectful populations organized in good political parties, in which open borders, free migration, multi-culturalism, humanitarianism, diversity, identity politics, intersectionality, racial harmony, radical feminism, development aid, democracy promotion, freedom of expression, political correctness, individualism, communitarianism, human rights, the Woke movement, and the Me Too movement will all flourish happily together amid the rising waters. These may be valuable goals in themselves, but they should be treated as entirely separate from the issue of limiting climate change.

There is an increasing and justified effort among Greens to re-label "climate change" as "climate emergency."[34] Well, the first things that get tossed out in a real emergency are luxuries. The Greens are declaring an emergency while clinging with fanatical determination to whole suitcases packed with what are effectively just ideological luxuries. If we cannot overcome our differences and take serious action to limit climate change and build national solidarity and resilience, we will be very lucky indeed if much of this agenda survives even into the second half of this century, let alone into the next one. The prospect of a failure to limit climate change has led to predictions that the 22nd century will be "the century from hell."[35] The population of hell is of course famously diverse; but if they have human rights, Satan has yet to be informed.

I

The Threat to States

There was a desert wind blowing that night. It was one of those hot dry Santa Anas that come down through the mountain passes and curl your hair and make your nerves jump and your skin itch. On nights like that every booze party ends in a fight. Meek little wives feel the edge of the carving knife and study their husbands' necks.

—Raymond Chandler, *Red Wind*[1]

The Bloomsbury intellectual, with his highbrow snigger, is as out of date as the cavalry colonel. A modern nation cannot afford either of them. Patriotism and intelligence will have to come together again. It is the fact that we are fighting a war, and a very peculiar kind of war, that may make this possible.

—George Orwell, *The Lion and the Unicorn*[2]

WHEN IT COMES TO assessing the risks of climate change, we really ought to look in a mirror, because the risk is us. Of course, major uncertainties remain in the science of climate change, above all concerning the timing and speed of particular effects; but by far the greatest uncertainty concerning predictions comes from the open question of how much humanity will do, or not do, over the next decades. What happens is up to us, and above all to the six major economies (counting the European Union [EU] as one economy) which between them account for almost two thirds of emissions and (given the growth of the

Indian economy and the huge scale of emissions from China) are likely to continue to do so for a considerable time.[3]

There is simply no more serious debate among expert scientists as to the cause and nature of anthropogenic climate change, though speed and extent are not certain.[4] That is why in this book I have not bothered to set out yet again the basic science. As with the approach of all sensible people to such issues, I accept the overwhelming scientific consensus unless or until I see a strong movement of genuine experts emerge to challenge it—and there is none.[5]

An Existential Danger

This authoritative scientific consensus states that if we remain on our present course, the global climate will warm by more than 3 degrees by the end of this century, possibly rising to more than 5 degrees if emissions continue to increase.[6] Even the commitments made under the 2016 Paris Agreement would probably only limit the rise in temperatures by 2100 to an extremely dangerous 3.2 degrees—and as of 2018–19 no major country was meeting its commitments, and annual CO_2 emissions were the highest ever.[7] Some experts like William Nordhaus have warned that even 5 degrees is an underestimate of the "worst case" scenario if emissions continue to rise.[8]

A rise of 3.2 degrees would not only bring disastrous results in itself but would also involve an extremely high risk of creating tipping points—like the melting of the Greenland icecap or the release of vast amounts of methane from the Arctic permafrost—that would lead to positive feedback loops and trigger runaway climate change.[9] If this occurs, it will eventually rise to levels that will create an "extinction event" like those of previous aeons, leading to the elimination of most species on earth including by far the greater part of humanity.[10] Once a tipping point has been passed, it cannot be reversed by any means that are now available to us or are likely to be available to us in the future.[11] If in consequence temperatures rise by 5 degrees over a period of decades, civilization will collapse. If they rise by 6 degrees, most of the planet will become uninhabitable by human beings.

Even if these apocalyptic consequences could be avoided, a rise of over 3 degrees would make certain the melting of the Greenland ice cap sheet, raising sea levels (over an uncertain period of time) by approximately 6 meters (20 feet) and drowning the world's coastal cities. It is highly probable that the Antarctic ice sheets would follow. The last time the world was 4 degrees hotter than now, sea levels were 260 feet higher.[12] For that matter, even a 2-degree rise would return global temperatures to the level of three million years ago—when sea levels were around 25 meters higher than they are now; and even at 2 degrees, weather patterns will be severely disrupted, increasing hunger across large parts of the world; heat waves in northern countries will rise to levels presently seen only in the tropics, while parts of the tropics may become uninhabitable; hurricanes will increase greatly in number and intensity; and the ice sheets will begin to disintegrate.[13] So even if a rise of 3 degrees does not trigger runaway climate change, the probable damage to the world economy has been estimated to be in the hundreds of trillions of dollars.[14]

No existing political system could survive economic decline on that scale. The difference between developed and developing states would be erased. This is why the Intergovernmental Panel on Climate Change (IPCC) now sets a rise of 1.5 degrees as the margin of safety—though according to the Mercator Research Institute (Berlin), the world already passed the point when that was realistically possible through reducing emissions in September 2018.[15]

Unless or until rising temperatures create a catastrophic tipping point like the melting of the Greenland ice cap, however, the effects of climate change on societies in the developed world will be incremental and will feed into other problems, strains, and conflicts including migration and economic dislocation. Disasters like hurricanes, droughts, and wildfires will multiply, but in the West, for a few decades at least, their direct effects will not be such as to endanger the survival of states.

But this does not mean that the effects of climate change will be of secondary importance. Precisely because states and societies are already facing a growing set of challenges, they cannot afford to suffer severe effects of climate change as well. This is the crucial thing about climate change in the medium term. It will feed into and exacerbate most other existing social, economic, health, and political problems—just as it will also feed into all other ecological problems, from mass extinction through deforestation to the acidification of the oceans.

The effects will be particularly bad in the developing world. Indeed, as I shall argue below, the Indian economy may well begin to decline by mid-century, with severe knock-on effects for the rest of the world. The chances that international aid will prevent such a decline are zero. This is the answer to those like Bjorn Lomborg who argue that we would be better off just leaving climate change to future generations because they will be richer than we are.[16]

I was struck in this regard by the latest book by Ian Bremmer, *Us vs. Them*, which paints a deeply worrying picture of the effects on Western societies of globalization, automation, immigration, and growing inequality. Nonetheless, at the end, he manages to extract at least some hopeful possibilities for the future. But Bremmer's book mentions climate change only once. Factor in the effects of climate change as well, and Bremmer's grim scenarios become even grimmer, and his hopeful ones a great deal less hopeful. The same is true of Paul Collier's impressive work on the future of capitalism, which mentions climate change only three times in 231 pages, or Dani Rodrik's equally penetrating work on democracy and the future of the world economy, which devotes only two pages to it in a discussion of "global commons."[17]

Rather than Robert Kaplan's vision of a "bifurcated world"—a secure and stable West threatened by "anarchy" elsewhere—what we are confronting is a global set of challenges to *all* existing states, albeit with different gravity and speed in different areas.[18] Of these threats, the single greatest one is climate change. These crises in turn will make it less possible to create political consensus behind action to limit climate change. Eventually, these effects will become so obvious and damaging that everyone will recognize the need for action; but by then social and political disintegration may have reached a point where no democratically agreed on consensus is possible, and states weaken to the point that they are incapable of effective action.

A recognition of the extent of danger to states leads to a recognition of climate change as a national security issue and of the need for states to take the lead. Individual actions, austerities, and sacrifices by environmentalists are morally valid and help in a small way to change public attitudes to climate change, but in the end states will have to take the lead, both in terms of actions and of shaping public consciousness.[19]

This is also the chief ground of my disagreement with Roger Scruton and certain other members of the conservative environmentalist camp, who prioritize the role of markets and limit the state to a role in research, and possibly taxation.[20] Scruton and I are at one in the importance that we attach to nations, and to what he calls *oikophilia*, or love of home. Oikophiles are essentially the Greek for what David Goodhart calls the "Somewheres." George Orwell could well be described as a progressive oikophile. I also share Scruton's debt to Edmund Burke, but with certain differences. Like Scruton, I do not think that Burke would have had much sympathy for the materialist frenzy that dominates contemporary society; but I believe that he would have recognized the need for this frenzy to be limited by state action for the common good. As he warned,

> Society cannot exist unless a controlling power upon will and appetite be placed somewhere, and the less of it there be within the more there must be without.[21]

Burke also made one great exception to his opposition to state involvement in the economy and preference for the local and the "little platoon": national defense; and indeed, in Burke's time the Royal Navy was the biggest industry in Britain. Today, US military spending is really a sort of (horribly inefficient) state technological-industrial plan that dare not speak its name in the presence of Republicans. Contemporary authors who have combined a desire for urgent action against climate change with reliance on capitalism have generally put state leadership and action at the heart of their vision.[22]

Of course, the local voluntary initiatives that Scruton praises—the little platoons of environmentalism—have a very important part to play as building blocks of environmental action and conservation, including energy conservation; but relying on them to limit climate change is a bit like arguing that Britain should have fought the Second World War by relying on the Home Guard.

Successful state action to limit carbon emissions requires not just determined state action but consistent action over a long period of time. This is something which for obvious reasons democracies find hard to achieve. The United States is the worst offender, with action against climate change at the federal level being virtually halted by the Republican victories in the presidential elections of 2000 and 2016.

Europeans like to congratulate themselves on their greater commitment to action against climate change; but in fact, their record is only comparatively less miserable. In Britain, investment in alternative energy dropped by almost 70 percent in 2016–2017, to £10.3 billion (less than one third of the UK defense budget, at a time when no direct military threat to the UK or its neighbors exists or is likely to develop) as a result of the government's abandonment of subsidies and introduction of taxation for alternative energy. In France, mass protests in 2018 forced the Macron administration to abandon even the simple measure of an increase in the diesel tax.

Worst of all, the limited damage caused to the Fukushima nuclear plant by the 2011 Tohuku earthquake and tsunami led to the abandonment of nuclear power by Germany—a country that has never suffered an earthquake or tsunami. This ensured continued reliance on coal and brought to an end what had been impressive progress to meet Germany's target of reducing the 1990 level of carbon emissions by 40 percent by 2020—a figure that Germany has now missed by a wide margin. So Greens too suffer from a severe inability accurately to calculate relative risks.

Climate Change, National Security, and the Military

In modern history, one of the chief areas where an appropriate consensus has been maintained over time has been in relation to national security; and the extension of national security to include climate change is only a further extension of a process seen since the end of the Cold War, by which the concept of security has been expanded to take in a range of new areas.[23] In China, India and elsewhere, concerns about the security of energy imports are already an important motive for state-led shifts to renewable energy. This process of "securitization" has been described and analyzed by the "Copenhagen School" of international relations theory, led by Barry Buzan, Ole Waever, and Jaap de Wilde.

They describe a process whereby a "speech act" by a recognized and authoritative national leader, institution, or party designates a particular threat to a particular society as a security threat.[24] Examples are the creation of the phrase "Cold War" in the late 1940s and the declaration of a "war on poverty" by President Lyndon Johnson. The creation of the phrase "climate emergency" is an attempt at such a speech act, but unfortunately by people who are not yet in a position to turn it into state action.

A speech act in the area of security requires the presentation of the danger concerned as an existential one and thereby removes it from the normal sphere of politics and policies. The threat is thereby placed in a special, exceptional category, backed by a national consensus and allowing the use of exceptional measures and the mobilization of national resources to meet it. That is why, as far as is realistically and constitutionally possible, the US military needs to throw its full weight behind the Green New Deal. In the words of a report by the Army War College "Army leadership must create a culture of environmental consciousness, stay ahead of societal demands for environmental stewardship and serve as a leader for the nation or it risks endangering the broad support it now enjoys. Cultural change is a senior leader responsibility."[25]

It is true that securitization has had negative effects in certain fields of policy: notably the way in which the United States turned the response to the terrorist attacks of 9/11/2001 into an all-engulfing "war on terror" and the even more inappropriate and damaging use of the word "war" in policies for crime and drugs. Liberal internationalists have condemned securitization out of their traditional hostility to national security and the nation state.[26] Many realists by contrast have argued essentially that real security threats remain those presented by states or by armed groups (most realists would now stretch a point to include terrorists) and that to extend the concept of security to other issues risks intellectual and policy confusion.

Nonetheless, in the case of climate change, securitization is appropriate and necessary: because this genuinely is an existential threat to all major states; because almost two generations after they began, efforts to tackle this issue through normal "politicization" and political mobilization (as advocated by most Green parties and movements) have failed; and because the mobilization of necessary resources and will does in fact have close analogies to the efforts required during war or at least acute armed competition.[27]

As Marc Levy has written,

> A threat to national security is an action or a sequence of events that (1) threatens drastically and over a relatively brief period of time to degrade the quality of life for the inhabitants of a state, or (2) threatens significantly to narrow the range of policy choices available to a state or to private, nongovernmental entities (persons, groups, corporations) within the state. . . . Taken all together, these effects [of climate change] would constitute a security risk if they threatened such a severe upheaval to the domestic economy that Americans would suffer greater hardship than we as a society consider tolerable.[28]

Climate change certainly threatens that. And after all, the point of any security policy, including a "classical" one, is the defense of an existing state and society and their institutions. Arguments over this are in the end nothing more than quibbling over words.[29]

It is striking in this regard that the two great developing countries that have been the most successful in developing renewable energy, China and India, both have strong and obvious security reasons for doing so. Both are heavily dependent on imported fossil fuels. Both therefore have reason to fear not just the effects on their economy of a surge in oil and gas prices due to instability in the Middle East, but also possible blockade of their maritime trade routes. This security motive does much to explain why India has done so much more to develop renewables than some much wealthier countries.[30]

As to the physical harm done by climate change to the lives of citizens (which is, or should be, at the core of legitimate definitions of national security, the direct effects of heat alone will kill far more people than all but the greatest wars. Even before climate change really kicked in, the European heat wave of 2003 killed some 35,000 Europeans—more casualties than those of France in the Algerian war lasting eight years. The Russian heat wave of 2010 killed around 55,000 people—twice as many Russians as died during the 10-year-long Soviet intervention in Afghanistan.[31] The years 2018 and 2019 saw record-breaking heat waves in Europe, Australia, and Canada. Australia in 2019–20 experienced the most devastating forest fires in its history, with dozens of lives lost and thousands of homes destroyed. As such heat waves continue and intensify, wealthy northern societies will have to introduce air conditioning on an enormous scale and at enormous expense. This will require considerably increased consumption of electricity, which—unless provided by renewables—will drive warming still further.

Moreover, through most of recorded history, security institutions have occupied themselves with a range of threats other than that of direct military attack by other states. These threats have included ideological threats to the ruling system and its political or religious ideology; and internal social disorder, whether political, ethnic, or criminal. Throughout the Cold War, the threat of the USSR and the West to each other was as much ideological, cultural, and economic as it was military, and the West eventually won not on the battlefield but on the field of ideology backed by economic success. Today, Western security elites are obsessed not with the threat of a direct military attack by Russia (whatever they may sometimes pretend in public for the sake of military budgets) but with the belief that Russia is subverting Western democratic processes in collusion with certain Western political forces.

As the next chapter will argue, if we continue on our present trajectory, then long before the direct physical effects of climate change become truly catastrophic, the indirect effects will combine with other social and economic strains to produce acute social and political disruption—almost certainly on a scale beyond the capacity of police forces to contain. Militaries will therefore be drawn inexorably into domestic crowd control and repression, which is a prospect that most soldiers view with absolute horror.

Even in Pakistan, where the military has so often taken power, the generals dread the prospect of using troops for this purpose in the regions from which the soldiers are themselves drawn—hence in part their long hesitation about confronting the Pakistani Taliban in the Pashtun areas of the country. If militaries are to avoid becoming turned into this hated role as armed police, they need to do everything they can to help prevent climate change and other developments that will make such a role inevitable.

It is vitally important to enlist national security establishments in the struggle against climate change for several reasons. Chief among them is the military's potential role as a bridge to those sections of the population that instinctively reject action against climate change on the basis of their political culture. Across most of the world, including the Western democracies, the military is the single most important institution when it comes to mobilizing the forces of nationalism behind climate change action.

In recent decades, attitudes to the issue of climate change among Republican Party supporters have moved away from even a pretense of considering the evidence and toward rejection of the issue on instinctive grounds: a belief that such rejection is part of what supposedly distinguishes Republicans from supposedly metropolitan, atheistic, decadent, unpatriotic cultural liberals. Not "We are not *convinced* by the evidence of climate change" but "We aren't *the kind of people* who believe in climate change."[32] However, while the conservative sections of the US electorate deeply distrust "experts," they make an exception for the military in their role as experts on national security. Climate change activists have debated the merits of taking an optimistic or pessimistic line in campaigns to educate the public on climate change and what can be done about it, but in terms of effectiveness, this largely misses the point. Rather, "Communication that affirms the sense of self and basic worldviews held by the audience has been shown to create a greater openness to risk information."[33]

In the United States and India, where denial of climate change is rooted partly in religious superstition, the modernity of military thinking can play a helpful role. Modern military establishments are by their nature modern in a way that political establishments do not need to be. As a Pakistani Air Force officer once told me, "There will always be a limit on fundamentalism in my service because to run a modern air force, you have to accept modern science."

The Calculation of Risk

Another contribution of the military to thinking about climate change is their approach to the calculation of risk and the prioritization of risks. Climate change deniers like to call for absolute scientific certainty before they are prepared to take action—a guarantee that action will come far too late. People who accept the reality of anthropogenic climate change in principle but oppose radical action also cite a lack of certainty. Even most climate change models, and economic assessments of the impact of climate change, omit the risk of catastrophic releases of methane from the Arctic permafrost and sea beds, because by their nature these releases cannot be quantified in advance.[34]

Yet this is generally recognized by experts as representing the greatest single possibility of triggering "runaway climate change" and further rises so high that civilization or even the existence of the human race itself would be threatened. The risk that the dying out of forests will contribute to climate change is also largely omitted from these models. The modeling of the economic consequences of climate change is even narrower, leading in some cases to absurd assumptions about modern civilization's capacity to survive increases in temperature to levels not seen for hundreds of millions of years.[35]

No soldier or military analyst thinks about threats in this way. The military operates on the basis not of certainties but of *risks*, the scale of risks and the balance between different risks, and there is a desperate need that they should extend this to the field of climate change:

> As abrupt changes and surprises do not lend themselves well to estimations of "most likely" probabilities (otherwise they would not be surprising), climate security assessments often therefore also focus more on what is possible than on what is probable. Military planning does take into account probable risks, but very often contingency planning is also made for events that are of unknown probability, yet entail severe consequences. . . . The emphasis is on responding to uncertainty, rather than on waiting for uncertainty to disappear.[36]

If the attitude to modeling of many economists and the attitude to risk of the climate change deniers were transposed to other areas of national security, then we would have to wait until there was a certainty that terrorists would acquire nuclear weapons before taking action to prevent them from doing so, or to wait till there was a certainty that Russia would invade the Baltic States before deploying forces to deter Moscow from doing so—by which time it would be much too late.

Awareness of the risks of climate change is now widespread in political and economic establishments. In 2019, the "Global Risks Report" of the World Economic Forum put environmental threats including climate change and extreme weather events (by now essentially the same thing) in three of its top five risks for likelihood and four out of five for impact.[37]

The risks come in two categories. The first concern the effects of climate change that we can already observe and that we can expect with near certainty to worsen in the decades to come: heat waves, drought, floods, increased levels of disease. Here, the imperative for national security experts is to compare these certain or highly probable effects with the actual or probable damage done to state interests by rival states, and decide which is the threat to prioritize. The second category concerns future damage that is not certain but that would be so catastrophic—involving the destruction of the nations that militaries are sworn to defend—that even the possibility of them should be enough to mobilize militaries in response.

Militaries by their very nature have to plan for the possibility of worst-case scenarios. From my interaction with them over the years, I can attest that Western security establishments often lean too heavily in the direction of paranoia, especially as far as Russia is concerned. Thus the Swedish government's warnings to its population of a possible Russian invasion ignore the existence of a small intervening geographical feature called the Baltic Sea, quite apart from a small military one called the US Navy.[38] Equally, however, to ignore the Russian threat altogether and to disarm as a result would be criminally irresponsible.

Barring a full-scale nuclear exchange between the great powers, no security threat in the world today comes anywhere near to matching the threat posed by climate change to existing states; nor is the damage being done by most states to each other remotely comparable to the damage *already* being done by climate change. Does the Chinese occupation of barely habitable sandbanks in the South China Sea threaten to kill thousands of American civilians through heatstroke every year? Or Russian occupation of parts of a worthless coal-mining region in Eastern Ukraine threaten to kill millions of Americans by tropical diseases over the next century? Seriously?[39] Yet the United States and the other great powers spend enormous amounts of attention and treasure in confronting what by comparison with climate change are minor threats.

China's claims to the Spratly Islands are certainly a violation of international law and challenge the United States' aspirations for unilateral hegemony in East Asia (which in any case Washington has never fully possessed). A glance at the map, however, should be enough to demonstrate that they are not a threat to international trade, on which China is even more dependent than the United States. Even a very much weaker US Navy, especially in alliance with India, would retain the ability to cut off China's maritime trade in response to any such Chinese move. And as noted, while the US military worries about the vulnerability of its own Pacific bases to rising sea levels, the Chinese should be far more worried.[40]

The central historical purpose of the militaries of the great powers has been to plan for a worst-case scenario—namely, a war between them—which has not happened since 1945. It is also in fact very unlikely to happen, given the balance of nuclear terror and other factors. At the same time, for militaries to ignore this risk in their planning would be insane, as long as the defense of their nations remains their core duty.

Planning for this worst-case scenario is of course also essential to deterring enemies and so making sure that such wars do not in fact occur. To defend these nations against the much greater and more real threat of climate change, it is necessary that states' understanding of risk be changed from that of their economists to that of their generals, and that the efforts that militaries currently expend to persuade elites and populations to pay much higher taxes for military defense be at least partially transferred to action against climate change.

The internal divisions in US society and politics concerning climate change are obviously serious barriers to the security establishment's playing a bigger role—as witnessed by the Trump administration's National Security Strategy of December 2017, which ignored climate change altogether.[41] However, the sheer scale of the threat to the security of the country means that the US military has an institutional and patriotic duty to instruct Americans concerning this threat, just as it has instructed them in the past on other threats falling within the military's sphere of competence.

Like other militaries, the US armed forces have frequently been involved in disaster relief operations at home and abroad. Unlike most other Western militaries, the US Army is also heavily invested in flood prevention and river management through the Corps of Engineers, which has always been the main US institution dealing with these areas.

As the effects of climate change increase, other militaries will have to imitate the US and Chinese examples and greatly increase their role in disaster prevention and relief, at the expense of other commitments.

The US Army Corps of Engineers will emerge in the future as the most important branch of the US armed forces; not a wholly new development, since historically an unusual proportion of American senior officers served for part of their careers in the Engineers (including Robert E. Lee, George Meade, George McLellan, and Douglas MacArthur). Since it was created in 1824, the Corps has been tasked with river management. In the course of the 20th century, and especially after the great Mississippi River floods of 1927 (causing damage which in today's terms would be well over one trillion dollars), it became responsible for the creation and maintenance of the world's largest system of levees and flood diversion canals. It is also responsible for controlling water pollution, restoring and protecting ecosystems including coastal wetlands, and maintaining the deepwater ports that handle more than two thirds of US imports—all of them tasks that are gravely endangered by climate change.[42] In the context of thinking about a Green New Deal, it is important to remember that the Corps was central to several of the original New Deal's key infrastructure projects, including the Tennessee Valley Authority.

In China, successful flood control and water management have been essential to the legitimacy of the state and have required conscription on a level equivalent to war. In the United States and other Western countries, the military has also been deployed whenever there has been a serious natural disaster; and in the future, governments and militaries will increasingly be judged above all by their competence in these areas.

In the States, this task also has an additional aspect: racial justice and the preservation of racial harmony. In the United States, floods have disproportionately affected the southern states; and the black population has both suffered disproportionately and been grossly discriminated against in the aftermath—something that remained true during the devastation of New Orleans by Hurricane Katrina in 2005. This has naturally contributed to the alienation of blacks from the US political system.

On the other hand, as Chapter 5 will argue, to make protection of minorities central to programs against climate change is a certain way to make many whites vote for the Republicans and thereby make serious state action much more difficult. Since the 1950s, the US military has emerged as a genuine force for racial integration and harmony in the United States. In the future, this will also need to be part of its reaction to floods and other impacts of climate change—not by rhetoric but in its visibly impartial and effective actions on the ground.

National Security and the Dead Hand of Tradition

When it comes to anthropogenic climate change, as of 2019 some military establishments talk the talk, but none of them really walk the walk. Trapped by their inherited structures, attitudes, and interests, they have failed to examine the correct balance between the different threats facing their countries. These residual elites came into being as a product of one set of historical circumstances and in response to one set of challenges but are proving incapable of changing to meet new and different dangers.[43] The same is true, by the way, of the greater part of the economics profession.

All national security establishments were created to meet the classical security threats of external invasion and domestic rebellion. Western ones evolved during the Cold War to meet the combined military and ideological threat of Soviet communism. Very little in their experience and structures equips them to think seriously about a completely new threat like climate change—especially since its worse impacts will hit far beyond the timescale of the usual military scenarios. Sometimes they simply cannot even recognize the existence of these challenges, since to do so would be to risk admitting their own redundancy.

There are significant and honorable exceptions to this pattern, such as the American Security Project and the Center for Climate and Security, both of which have long lists of distinguished retired generals and admirals on their boards.[44] In March 2019, 37 retired US generals and admirals (including General Stanley McChrystal, former commander in chief in Afghanistan) joined other national security figures (including a few Republicans like Chuck Hagel) in signing a letter to President Trump protesting politically driven assessments of climate change by the National Security Council. They ended, "We spent our careers pledged to protect the United States from all threats, including climate change."[45]

Unfortunately, such statements are invariably made by retired officers. Serving generals and admirals have been far more circumspect, even under Democratic administrations—while showing no such circumspection in talking up traditional security threats. The same has been true in the UK and Europe. In December 2018, the chief of the British armed forces, General Sir Nicholas Carter, included Russia, China, migration, and populist nationalism as "existential threats" to the United Kingdom—but he did not mention climate change. A simple Google search turns up statements by other serving and retired military chiefs including General Mark Carleton-Smith, Air Chief Marshal Peach, and Admiral Lord West warning of the Russian threat. A similar search finds no statements by these figures on climate change. Not surprisingly, therefore, the media in response highlights traditional security threats and not those of climate change.[46]

Even Western think tanks specializing in foreign and security policy, though they take climate change more seriously, generally place it in a separate box from security issues. They thereby ensure that most security experts will never read their reports. I have personally experienced again and again how experts on Pakistan who focus on short-term security threats to that country completely ignore the existential long-term threat posed by the combination of climate change, population growth, and the country's nightmarish water situation. Over the past generation, it has been entirely possible to conduct a prominent career in security studies and policy advice without mentioning climate change at all in your publications.

So it is not that national security establishments of the great powers have completely ignored the threat of climate change. With the exception of Russia, all have publicly recognized its existence.[47] In the United States, the Pentagon dedicated part of its Quadrennial Review of 2010 to the subject and has continued mentioning it even as the Trump administration denied its existence. The US Defense Authorization Act for 2018–19 submitted to Congress by the Pentagon read in part,

> As global temperatures rise, droughts and famines can lead to more failed states, which are breeding grounds of extremist and terrorist

> organizations. . . . A three-foot rise in sea levels will threaten the operations of more than 128 United States military sites, and it is possible that many of these at-risk bases could be submerged in the coming years.[48]

The problem is one of balance, publicity, and influence. Compared to "classical" security threats (among which terrorism must now be included, since 9/11 provided the incontestable event that made earlier disregard for the topic's seriousness obsolete), the subject of climate change occupies only a small part of military statements and attention, or the influence that the military brings to bear on Congress, the media, and public opinion. Climate change had almost no presence in the Quadrennial Defense Review (QDR) of 2018, which declared instead that "the central challenge to U.S. prosperity and security is the re-emergence of long-term, strategic competition by what the National Security Strategy classifies as revisionist powers." In recent years, the vast majority of statements about risks to the country made by senior US military figures have been concerned in the first place with traditional great power threats, followed by terrorism.[49] As Chad Michael Briggs of the US Air Force University warns, "Use of the term 'climate change' in policy documents does not mean that associated risk assessments have been mainstreamed into [US] military planning."[50]

Or as the Rand Corporation stated concerning the North Atlantic Treaty Organization (NATO),

> In the case of nuclear weapons, terrorism, and cyber issues, each offers more uncertainty than climate change. However, vast amounts of resources are dedicated to the sponsoring of research, understanding the threat, and the preparations for potential consequences. The contrary is true for the potential security impact of climate change. . . . The lack of engagement at NATO headquarters on this point is more appropriate for the management of a tolerable or acceptable risk, while the literature suggests that climate change presents risks that likely won't be tolerable or acceptable.[51]

The 2008 National Security Strategy of the United Kingdom declared roundly that "climate change is potentially the greatest challenge to global stability and security, and therefore to national security."[52]

The previous year, British foreign secretary Margaret Beckett had declared that "achieving climate security must be at the core of foreign policy."[53] In the decade since then, however, British governments have not come close to the kinds of actions that they would have taken if they had truly assimilated the idea of climate change as a principal challenge to national security (though to their credit they have commissioned some of the most important research into climate change, including the Stern Report).

Instead, the security agenda and the attention and expenditure associated with it were frittered away on "traditional challenges": a war in Afghanistan which by 2008 was already effectively lost and in which Britain achieved nothing; hysteria over minor post-imperial squabbles over disputed territories in the former USSR that had never been of the slightest interest to Britain; and panics whenever a Russian warship sailed past Britain, assiduously stoked by British admirals desperate to preserve their shrinking budgets. The attention paid by the British media to these issues usually dwarfed that given to climate change. Even when it comes to the warming of the Arctic, Western security experts concentrate on the minor threat of increased Russian and Chinese presence there and not the existential threat of methane release and ice cap melting.[54]

The same has been true of the UK budget. In these years the UK spent £8 billion ($10.3 billion) on the war in Iraq, £21 billion ($27 billion) on the war in Afghanistan, £6 billion, or $7.2 billion (not counting planes) on the new British aircraft carriers, and planned to allocate £31 billion ($40 billion) for Britain's new nuclear deterrent. By contrast, in 2016–17 spending on alternative energy dropped by almost two thirds to around £7.5 billion ($9.7 billion).[55] This represents a completely false set of priorities concerning the vital interests of the United Kingdom.

If the scientific predictions concerning climate change and its impact are correct, then a hundred years from now, most of the preoccupations of today's security and economic establishments are going to seem not just irresponsible but senseless. As noted in the preface, the military bases that China is building on reefs and sandbanks in the South China Sea are a particularly striking case in point. Immense Chinese efforts are going into the creation of these bases and their accompanying claims to sovereignty over the sea, and these actions have raised regional tensions and cost China badly in diplomatic terms.

Meanwhile, the United States has deployed considerable military forces to the region in response to these moves and is running considerable risks—even potentially of war—in resisting China's claims. Yet if they had truly assimilated this recognition into their strategy as a whole, they ought to be encouraging China to pour concrete, sand, and money into bases that will either be under water 100 years from now, or will at the very least be repeatedly rendered militarily useless and have to be rebuilt often after typhoons. And of course, the Chinese state, which on balance takes climate change much more seriously as a threat than does the United States, can be blamed even more for this dangerous and pointless effort.

Robert James Woolsey, former head of the Central Intelligence Agency (CIA) between 1993 and 1995 is one of the few leading members of the US security establishment to have talked about climate change as a major security threat not just to the United States itself but to American leadership in the world. As he has written,

> In a world that sees two metre sea level rise, with continued flooding ahead, it will take extraordinary effort for the United States, or indeed any country, to look beyond its own salvation. All of the ways in which countries have dealt with natural disasters in the past . . . could come together in one conflagration: rage at government's inability to deal with the abrupt and unpredictable crises; religious fervour, perhaps even a rise in millennial end-of-days cults; hostility and violence towards migrants and minority groups, at a time of demographic change and increased global migration; and intra- and interstate conflict over resources, especially food and fresh water.[56]

Yet in the eleven years since he wrote this, rather than continuing to drum home this message at every public opportunity, he has instead reverted to previous preoccupations and dwelt obsessively on the supposed threats to the United States from China and Russia.

Mr. Woolsey and other US establishment advocates of a new Cold War with Russia and China constantly use old Cold War language of an ideological struggle between Western democracy and authoritarianism to support their positions; but the issue of climate change significantly blurs the moral distinction between democratic and authoritarian systems—or at least, that is likely to be the view of future historians. Relative to their per capita incomes, authoritarian China can be classed with liberal democracies like Denmark when it comes to current determined action to reduce carbon emissions.[57]

Among the democracies, not just the United States but Canada too can be classed with semi-authoritarian Russia as the greatest offenders against climate change action—not on ideological grounds but simply out of a desire for economic gain from the exploitation of fossil fuels. Worst of all (in terms of the relationship between per capita income, per capita contribution to CO_2 emissions, unused opportunities for alternative energy, and already obvious severe damage) has been democratic Australia. In February 2019, its Liberal-National government called emergency talks to discuss a combination of devastating floods, heat waves, drought, and forest fires—while still not discussing climate change because of a refusal to move away from dependence on coal and coal exports—and was re-elected in May 2019![58] The Australian right is also obsessed with the threat of Asian migration, which climate change can only drastically worsen.

In opposing implementation of the 2016 Paris Agreement, the US Trump administration, Russia, Canada, and Australia were on one side, and China and the European Union on the other—which means that the most important issue facing humanity cuts clear across the supposed "new Cold War" alignments. Truly, one can echo the words of the great theologian and Christian Realist thinker Reinhold Niebuhr: "There is only one empirically provable element in Christian theology, namely, that 'All have sinned and fallen short of the glory of God.' "[59]

The Impact of Climate Change on the Global North

If the catastrophic scenarios of a 5 or 6 percent rise in temperatures come to pass, then all existing states will be overwhelmed and by far the greater part of the human race will be doomed. When it comes to the scenario of a rise in the 2- to 3- degree range, however, the direct effects on the major states will differ very considerably.[60] For most of the developed states of the world, the direct physical consequences of a rise in temperatures of up to 3 degrees over the next century are likely to be exceptionally unpleasant by the Western standards of the past 70 years but just about manageable in purely physical terms. Thus the HSBC report of 2018 on vulnerability to climate change puts Russia in 66th place, the United States at 39th, and European countries stretching from the 30s to the 60s.[61]

The greater risk to them will be indirect, above all from increased migration as the century progresses (see Chapter 2). Among the neighbors of the United States, Mexico stands eighth in the world on HSBC's vulnerability index, and Colombia seventh.[62] Almost half the population of Central America lives below the poverty line, water shortages already affect large areas, and hurricanes have demonstrated a capacity to knock back economic development by years or even decades.[63] Moreover, as we have seen repeatedly over the past two centuries, the malign racial, socioeconomic, and political legacies of Spanish colonial rule combine to make the region exceptionally violent, oppressive, and politically unstable.

Over the past generation, US administrations replaced the military interventions of the past with indifference toward their southern neighbors, inadequately veiled by the false promise that the North America Free Trade Agreement (NAFTA) would transform the region economically. The failure of NAFTA to do so has led to the continued flow of illegal migration to the United States, which has done so much to infuriate sections of the white population and to elect Donald Trump.[64]

US military officers, security analysts, and sympathetic journalists worry incessantly whether the failure to intervene militarily in Syria or "stay the course" in Afghanistan will damage US "credibility." Nothing has so damaged US credibility in the world as the decay of its domestic political system. The poverty and despair of America's southern neighbors contribute greatly to that decay; and climate change will contribute enormously to their suffering.[65] As Elizabeth Warren has proposed, one important part of a Green New Deal in the United States should therefore be increased aid to develop Central America and strengthen its resilience in the face of climate change—something that can be justified to recalcitrant voters and their representatives by the need to limit migration.

The direct damage to the West will also be bad enough. Deaths as a result of heat waves will soar, possibly reaching the tens of thousands annually and vastly outpacing the casualties of the largest ones in the past. Incidents like the Chicago heat wave of 1995, which killed 739 people in five days, will become a regular occurrence. Tropical diseases will spread northward (tick-borne Lyme disease has spread enormously in the United States over the past 20 years, apparently as a result of rising temperatures).[66]

In this country, a disproportionate number of the greatest cities, including New York, Los Angeles, Miami, and Seattle, are in low-lying coastal areas in acute danger from a rise in sea levels and from worsening storms.[67] It is important to remember that long before places disappear permanently under water, repeated flooding will make them uninhabitable. Thus, it is estimated that by 2045, some 300,000 US homes with a total current value of around $117 billion will be uninhabitable due to flooding, while by the end of the century homes and businesses currently worth more than $1 trillion (not including infrastructure) are likely to be at risk.[68]

The great breadbasket of the Midwest is highly susceptible to drought and consequent increased soil erosion;[69] large areas of the country are vulnerable to the kinds of wildfires that hit California in 2018; and the naturally arid or semi-arid Southwest (including the huge population center of Southern California) is heavily dependent on water supplies from endangered snowmelt in the Rocky Mountains.

Given the radical decline in the great underground aquifers of the western United States, states and communities are going to have to make some politically and socially wrenching choices between the needs of cities and the needs of agriculture. Los Angeles, San Diego, and the Bay Area in California will have to raise vast sums to pay for the desalination of seawater. Increasingly severe limits on consumption will have to be put in place, putting an end to the lawn, swimming pool, and golf club culture that has attracted so many middle-class people to the region in the first place—even as these are also scorched by increasingly murderous heat waves. In fact, we are likely to see a reversal of the migration of the past decades to the Southwest from the rust belt of the Midwest, bringing an end after only a few decades to the huge urban developments in the region. One US study by Dr. Jesse

Keenan of the Harvard School of Design has suggested that Duluth, Minnesota, will be the ideal US destination city in the future—not a thought that would have occurred to anyone 20 years ago.[70]

In Europe, the most dramatic direct effects of 2 to 3 degrees of global warming will be seen in the Mediterranean, where the summer is predicted to last for an additional month, heat waves (with temperatures over 35 degrees) to be extended by more than a month, and rainfall to decrease by up to 20 percent. The result will be severe damage to existing agriculture, the radical transformation of ecosystems toward semi-arid conditions of the type now common in countries such as Pakistan and Australia, a steep decline in tourism, and greatly increased wildfires. Runaway climate change would lead to the complete desertification of the region.

The start of these effects is already here, as demonstrated by the repeated heat waves of recent years and the unprecedented forest fires of 2018 in Greece and Portugal and of 2019 in Siberia. Even with 2 to 3 degrees of warming, they are bound to get much worse. Moreover, these countries of southern Europe are also those that will be asked to accommodate the largest number of migrants from the even worse-affected countries on the southern shores of the Mediterranean. Does Russia's intervention in Ukraine threaten to turn Italy and Spain into an extension of the Sahel? Seriously?

Northern Europe will be less affected but will find that the "extreme" summer temperatures of 2015–18 are now the norm, with periodic heat waves vastly in excess of that, leading to more mass casualties like those experienced by France during the heat wave of 2003 and more wildfires like those affecting Sweden in 2018. Coastal areas in northern Europe will experience increased flooding. Cherished landscapes will be radically altered as vegetation dies or migrates northward and is replaced by new plants from farther south. Tropical diseases like dengue fever and malaria will spread to northern Europe.

In Russia, as in the United States and India, anthropogenic climate change denial begins at the very top, despite the fact that the average national temperature has risen by almost twice the global percentage over the past 30 years. Vladimir Putin has repeatedly expressed skepticism on the subject. Other parts of the Russian establishment do on the whole acknowledge the basic science, but they are highly resistant to taking serious action to reduce emissions. Russia has failed to meet even its very low goals for changeover to alternative sources of energy, and it joined with the United States and Saudi Arabia in 2018 to block full acceptance of the latest IPCC report calling for 1.5 degrees as the new limit for global warming.

The basic reason for these attitudes, as elsewhere, lies in the prioritization of economic development and calculations of basic national advantage and disadvantage: the cost and difficulty of the shift to alternative energy coupled (as in Australia and Canada) with the colossal benefits to the Russian economy of the production, exploitation, and export of fossil fuels. These priorities are shared by the population as well as the elites. In an opinion survey of June 2014, only 18 percent of respondents placed the environment among their chief concerns.[71] The government is deeply afraid of popular unrest resulting from lower growth, and is therefore highly resistant to anything that would further reduce growth.[72]

The difference in the case of Russia (as in Canada) is the widespread belief that climate change will on balance actually *benefit* the country. At bottom, this is due to the very human reason that in a country with Russia's traditional climate, people do not instinctively see greater warmth as a bad thing. However, they may also be right in purely physical and economic terms, as long as the rise in temperatures does not exceed 3 to 4 degrees Celsius.[73]

Like Canada, but very unlike Australia, Russia's northerly position and mostly distant and unpopulated coastal areas mean that in the short to medium term the negative effects of climate change will be limited. Of Russia's major cities, only St. Petersburg and Vladivostok are seriously menaced by rising sea levels in the medium term. Natural disasters like heat waves, forest fires, droughts, and storms, together with the spread of tropical diseases, will—or so it is believed—be accepted and absorbed by the famous toughness and resilience of the Russian people.

In addition, and most important, Russia's large agricultural production and small population relative to territory (factors it shares with Canada) means that it has a cushion against agricultural crises that is largely lacking in the EU, China, and Japan, and wholly lacking in South Asia. There is also a widespread assumption—both popular and official—that any negative effects on agriculture will be more than balanced by the warming of northern areas. Climate change has already brought about the opening of the Arctic seaway to commercial traffic, and Russians hope that it will allow the massive future extraction of gas and oil reserves from beneath the Arctic Ocean.[74]

Meanwhile, the drastic deterioration of relations with the United States and the European Union since the Ukraine crisis of 2014 have made the Russian elites even less willing to listen to lectures from the West on international responsibility. In these circumstances, the only way to appeal to the Russian establishment and population to take action against climate change is by an appeal to their nationalism and to the long-term national interests of Russia.

The threat of limited climate change in this regard stems from increased international disorder and international migration. As a great Eurasian state, Russia will be just as threatened as the EU and India by the collapse of other Asian states. Russia has run considerable risks and deployed considerable military power to save the Syrian state from collapse, fearing—quite rightly—the spread of Islamist extremism and terrorism that would result from a lawless state. Even limited climate change, however, is likely to encourage a long sequence of new Syrian civil wars to Russia's south. And of course, if global warming accelerates toward 5 or 6 degrees Celsius, then Russia itself will face the same existential threats as every state around the world.

China's Historical Experience

China is much more endangered than Russia because of the far greater size of its population, the concentration of that population along endangered seacoasts, the more precarious state of its agriculture and water supplies, the threat of drought, and the long-term threat to its great rivers from the melting of the Himalayan glaciers. The extraction of too much water from China's great rivers, coupled with the rise in sea levels, risks salt water creeping farther and farther inland, gradually crippling some of China's most fertile agricultural areas. The building of giant hydroelectric projects helps limit carbon emissions but does still further damage to water supplies downstream.[75]

Chinese water use for agriculture is efficient only compared to South Asia and Africa, its industry even less so. Water abundance in the south and scarcity in the north may create regional tensions between different regional factions of the Communist Party. In 2005, China's minister of water resources reportedly declared, "Fight for every drop of water or die. That is the challenge facing China."[76]

As climate change worsens, increases in wheat production in northern China as a result of increased temperatures risk being canceled out by drought, in a country where water shortages are already becoming a serious problem. In the south, rising temperatures threaten to reduce China's staple crop of rice. Should the Himalayan snowpack significantly diminish, China will face a critical agricultural crisis. Not only will its own exports cease, but it is probable that the enormous financial reserves of the Chinese state will allow it to suck in food from much of the developing world, increasing prices around the world and contributing to global shortages and unrest elsewhere.

The Chinese drought of 2010 led China to buy huge amounts of wheat on world markets. Coupled with droughts in Syria, Russia, Ukraine, and Argentina and torrential rains in Canada's wheat belt, this drove up bread prices and thereby contributed to the discontent of Middle Eastern populations that led to the "Arab Spring" the following year. This in turn has had vast and ongoing consequences for regional and global security, for the growth of Islamist militancy, and for US relations with Iran, Turkey, and Russia.[77]

In China in recent decades, poverty in the agricultural provinces of the interior has been greatly ameliorated by migration to the booming cities of the east and southeast coast. As sea levels rise and storms intensify, these areas will, however, themselves be increasingly menaced by flooding. China risks being caught in a giant demographic pincer, whereby tens of millions of people fleeing the drying interior run up against tens of millions of others fleeing the flooded coast.

A severe shortage of water for drinking, cooking, and bathing (like electricity cuts in South Asia in summer) radically diminishes the quality of life, irrespective of how many designer handbags or iPads people have been able to buy. I vividly remember as a journalist in Afghanistan the miseries of a lack of water: the lines of women trudging for miles in 40-degree heat with buckets of water precariously balanced on their shoulders and heads; the need to ration every cupful of water when on the march with the guerrillas, and the torments of thirst when the last cupful was gone. In both India and China, a radical increase in such shortages will negate the social benefits of economic growth and threaten political stability—even before they bring economic growth itself to an end.[78]

If the current rulers of China have taken climate change far more seriously as a threat than have their American or Russian equivalents (or the Europeans, for that matter, proportionate to incomes per capita), it is above all because of historical experience. Past natural disasters in China have been on a vastly larger scale than those in the United States, Russia, or western Europe and have had a far greater capacity to undermine the legitimacy of the state. The costliest ever natural disaster in US history in terms of lives lost was the hurricane of 1900 that overwhelmed Galveston, Texas, killing between 6,000 and 12,000 people. The most damaging European floods in modern history, those of 1953, killed 2,142 people in four countries.

The 1931 floods in China killed up to four *million* people (mostly indirectly, through malnutrition and disease) and displaced 14 million; and they were only one of six Chinese floods in the 20th century that killed more than 100,000 people.[79] There has never been a famine in US history, and those in Europe and Russia over the past century have been caused by war and state action. In the course of the 20th century, half a dozen Chinese famines due to natural causes resulted in an estimated 35 million dead. So the Chinese know what they have to fear.

According to one of the most ancient Chinese legends (but with clear origins in fact), following the Great Flood of Gun Yu, the greatest achievement of Emperor Yu the Great (ca. 2123–2025 BCE) was to create systems of dikes and canals to control the rivers and irrigate the surrounding lands. For 4,000 years since then, floods, droughts, and famines have been taken as a sign that the ruling dynasty had lost the "Mandate of Heaven"—in other words, its political legitimacy. And insofar as a failure by the state to mobilize enough people to repair the dikes and irrigation canals helped gravely worsen the effects of natural disasters, there was indeed a relationship to state weakness and incompetence.

Famines in turn led to outbreaks of peasant rebellion and the appearance of bandit armies, thereby undermining the state still further.[80] The Chinese famine of the late 870s CE led to a rebellion that helped destroy the T'ang Dynasty. The famine of 1630–31 is supposed to have done the same for the Ming; and the successive famines of the later 19th and early 20th centuries helped to fatally weaken the Manchus, as did the great flood of 1855 when the Yellow River changed its course by hundreds of miles, with dreadful human consequences.[81] This is a pattern that climate change threatens to replicate across much of the world.

Due to this historical awareness, the case of China illustrates how nationalism, state interest, and elite interest can provide the basis for a determined effort to combat climate change. The greater Chinese state efforts in this regard are certainly not due to altruism directed at humanity in general. The rulers of China do, however, have an exceptionally strong sense of having inherited an ancient state, which they are determined to pass on to their descendants; and it helps when trying to look 200 years into the future if you can look 4,000 years into the past.[82]

But while China is taking strong action against climate change proportionate to its wealth per capita, the sheer size and growth of its economy have meant that emissions have continued to rise regardless, albeit more slowly.[83] China has only promised to start reducing its total emissions as of 2030. Whether it can do so rapidly enough after that depends in part on alternative energy, into which it is pouring money at a far higher rate than the United States, especially given China's very heavy dependence on coal for electricity generation. China is the world's largest consumer of coal by far, and its continued increase (4.5 percent in 2018) contributes greatly to China's failure to meet its commitments under the Paris Agreement.[84] Coal burning accounts for more than 70 percent of China's CO_2 emissions. Whether the Chinese can really reduce that quickly enough seems highly questionable.

The greatest contribution to date that the Chinese state has made to reducing climate change emissions and environmental impact was its "One Child" policy, the ruthless limits on births imposed between 1979 and 2015, which the government estimates led to a population that was around 300 million (22 percent) lower than it would otherwise have been. Another truly radical development that the Chinese are already engaged in is the plan to make 50 percent of vehicles electric by 2025, and 100 percent by the 2030s, through a mixture of subsidies and sanctions. If they manage this (and there are enormous obstacles in the way), it would be a very important contribution to limiting carbon gas emissions and also a very convincing testimony to the ruthless efficiency of the Chinese system.[85] If, however, they fail radically to reduce their emissions by mid-century, then after India, they will be the worst sufferers among the major powers, to a degree that may bring down the Chinese state in the second half of this century, long before the direct global physical effects of climate change become truly apocalyptic.[86]

India and South Asia

The vast majority of reporting and analysis of security issues in South Asia and the Persian Gulf in recent years has related to traditional security threats: the war in Afghanistan, the threat of terrorism, the Kashmir dispute between India and Pakistan and the danger of nuclear war between them, US and Israeli threats to Iran, the geopolitical and religious rivalry between Saudi Arabia and Iran, the Saudi-led boycott of Qatar, and so on. Indian and Pakistani public opinion and the media have also focused overwhelmingly on security issues in South Asia, including the tensions between them over Kashmir and terrorism, rather than on climate change.[87]

Almost unnoticed by security institutions—including those in the South Asian countries themselves—have been two reports on the dangers of climate change by scientists of the Massachusetts Institute of Technology (MIT). These reports present evidence that across large areas of South Asia and the Gulf, by the last quarter of this century, climate change leading to extreme heat waves is likely to make it impossible for human beings to work in the open for much of the year. Barring enormous and enormously costly adaptation, agriculture across much of the region will also be severely damaged.[88]

A 2018 report by HSBC puts India first among large countries vulnerable to climate change, followed by Pakistan in second place and Bangladesh in fourth place. Pakistan also scored third from the bottom when it came to potential to respond to climate risks (Bangladesh was eighth from the bottom and India 10th).[89] In South Asia, the truly disastrous consequences of climate change will therefore begin to kick in at much lower levels of global warming than in most of the rest of the world.

The threat to human life from heat waves in South Asia comes from a combination of extreme heat with humidity, measured by a reading called the "wet bulb temperature," which measures the ability of moisture (including sweat) to evaporate. Humidity, which blocks this evaporation, means that even a relatively small rise in South Asian temperatures will be fatal. Thus a wet bulb temperature of only 35 degrees Celsius means death after a few hours because the human body cannot cool itself enough to survive.

According to the MIT study, if present trends continue, the proportion of the population exposed to wet bulb temperatures of 32 degrees Celsius—very close to the survivability threshold and extremely hazardous to health—will increase from 2 percent of the South Asian population at present to 70 percent. At these temperatures, sustained work outdoors will be impossible.

At sustained temperatures of over 40 degrees Celsius, rice cultivation becomes impossible. At present, these temperatures last only a few weeks. With climate change, they could last for months on end, wiping out the staple crop of much of the region. A foretaste was given by the collapse of the Sri Lankan rice harvest in 2017 due to drought.[90] The IPCC report of 2018 estimated that a further rise in temperatures of only 0.5 degree Celsius would reduce India's grain harvest per hectare by more than one sixth.[91]

So while the other great powers may be able to survive global warming if it remains within the 2- to 4-degree range, this is not true of India. Even at 1.5 degrees, the effects of climate change will be enough to endanger the stability of the countries of South Asia, which is home to almost a quarter of humanity, and also a quarter of the world's malnourished.[92]

The World Bank predicts that if we continue emissions at the rate of recent years, by 2050, in South Asia alone some 800 million people (around 35 percent of the probable population at that date) will see their living standards decline sharply as a result of climate change.[93] Think about this a bit. In 2050, Indian teenagers alive today will only be middle aged. It is not a distant prospect affecting generations yet to be born. At present, Indian governments worry deeply about whether Indian growth will be high enough to provide enough jobs for the millions of new people entering the job market each year, and the social and political consequences if it is not. Think what de-growth would mean. Admittedly, de-growth would bring down Indian carbon gas emissions—but at what a human cost!

This prediction makes nonsense of the argument that India needs to go on expanding its fossil fuel consumption in order to power economic growth. If we continue on the present path, Indians alive today will see economic growth go into reverse, with incalculable social and political consequences. The dream of "India Shining" would be over. The Indian elites have invested immense emotional commitment in the hope of India becoming a superpower on a par with the United States and China. On present climate trajectories, many of them will live to see that hope irretrievably collapse. If in addition the resulting suffering is very unequally distributed among Indian states and leads to mass migration within South Asia, it is hard to see how Indian democracy and the Indian Union itself will be able to survive.

According to an Indian government report, by 2030 India will have only half the water it needs to sustain existing levels of consumption, let alone higher ones due to economic growth. India's water consumption is predicted to be 1.2 billion cubic meters by 2030, 50 percent higher than the figure for 2012. That would be an extremely difficult level to sustain even without the effects of climate change.[94] In both India and Pakistan, water shortages are causing friction between upriver and downriver provinces. Should the water crisis become truly disastrous, these tensions have the capacity to spur both separatist movements wanting to secure water supplies and violent state reactions against them.

Moreover, even if economic development allows India to deal adequately with the direct effects of climate change, the country is likely to be overwhelmed by the collapse of the even more endangered neighboring states of Pakistan and Bangladesh.

Even a 2-meter rise in sea level by the end of this century has been estimated to displace up to 200 million people—and 2 meters is beginning to look like a conservative estimate. As Chapter 2 will argue, the resulting waves of migration from South Asia and other regions will in turn be the most dangerous indirect effect of 3 degrees or so of climate change on Russia and the West. In the succinct words of the headman of a south Indian village, "There is no water. Why should people stay here?"[95]

Many parts of South Asia are already experiencing severe water stress, and since 2000, drought in north India and Pakistan has largely wiped out the expected further gains from agricultural development. Drought, coupled with the commercialization of agriculture, is causing despair among India's smaller farmers (leading to a surge in rural suicides over the past two decades) and is fueling support for the Naxalite communist rebellion in central India.[96]

Even below 2 degrees of warming, climate change significantly increase the variability of the monsoon, leading to an increase in both floods and droughts.[97] A 2018 report by the official NITI Aayog institute in India stated that 600 million people (almost half the population) already suffer high to extreme water stress.[98]

As in Pakistan, these water shortages are not necessarily absolute. A very large element in the growing crisis is created by wastage and poor infrastructure. Indian farmers use twice as much water per ton of wheat as their equivalents in China and the United States, and Pakistani rates are even worse. Since irrigation for agriculture accounts for around 80 percent of Indian water use, the damage is colossal.[99] The problem could be greatly ameliorated by water-saving approaches (especially the use of micro-irrigation in agriculture), better repair and maintenance of canals and water supply systems, and better rainwater harvesting.

The performance of Indian states differs considerably in this regard, but unfortunately, almost half of Indian states fail to reach 50 percent of the possible score when it comes to water management, and more than half of the population lives in the worst performing states. As of 2018, only about 10 percent of India's irrigated land was watered with drip or sprinkler methods, compared to around 70 percent of China's. It has been estimated that in India's cities, 40 percent or more of piped water is lost through leaks. Figures for Pakistan are comparable or even worse.

So far, the radical drop in the water table in many areas as a result of over-use of tube wells has had very little effect in this regard (something which is also true in the southwestern United States). Across India, 21 major cities are expected to run out of groundwater altogether within the next few years. These include Bangalore, the heart of India's IT industry and a supposed showcase of Indian modernization.[100]

In Israel, the country that leads the world in water conservation, the most important contribution to this has been water pricing to push the population not to waste water. According to former Israeli water commissioner Shimon Tal, "For the few years before the price mechanism was used, we were in the middle of a terrible regional drought. . . . Then we used price as an incentive. Almost overnight, consumers found ways to save nearly *double* the amount of water they had saved because of our years-long education campaign. It turned out that price was the most effective incentive of all."[101]

Across the wider region, however, I have been told that the introduction of water pricing for agriculture is simply politically impossible—even as in cities, the breakdown of state water supplies has reduced many of the poor to buying their water from tankers or simply stealing water by breaking into the pipe system. The much stronger and more authoritarian Chinese state has also shied away from water pricing or taxation. For such painful measures to be introduced without increasing hostility to the state requires a strong sense in the population (not just the ruling elites) of common national danger and collective national will.[102]

Much of the literature on climate change and conflict has focused on the possibility of interstate wars over rivers, including between India, Pakistan, and China as increased water shortages and the melting of glaciers lead to worsening disputes over the sharing of the Indus and Brahmaputra Rivers. Though by no means absent, this threat may have been somewhat exaggerated. Precisely because these rivers are so vital, serious interference with them would be regarded as an existential threat. Pakistan has stated that serious Indian reductions in the flow of the Indus would be regarded as the equivalent of war; and presumably China, as Pakistan's ally, would respond in kind by reducing the flow of the Brahmaputra River to eastern India.

However, if water shortages become so severe that the choice for upstream states is water diversion or state collapse, then all bets would be off. It seems likely, though, that by the time this point is reached, a combination of internal stress and mass migration would already have brought the states concerned down in ruins.

Three out of India's four largest cities lie on the sea or estuaries, including Mumbai, the largest. This is also true for Karachi, Pakistan's largest city, and for Dhaka, the capital of Bangladesh. Even with greatly enhanced flood defenses, a sea-level rise of 1.5 meters (now widely predicted even if the rise in temperatures is kept to 2 degrees) would threaten as much as 40 percent of Mumbai and the homes of around 12 million people.[103] And this is on the assumption that climate change can be kept within these limits. Should a significant part of the Arctic and Antarctic ice sheets melt, Mumbai and the other coastal cities of South Asia will simply disappear.

In Bangladesh, it has been estimated that a rise in sea levels of 1 meter would render 17.5 percent of the country uninhabitable, while a rise of 10 meters would essentially destroy the country. By the end of the century, Bangladesh is predicted to have a population of 250 million at the absolute minimum (from 160 million at present). This means that even with a 1-meter rise, tens of millions of people will have to move. In addition, a rise of 1 meter in sea levels would almost entirely submerge the Sundarban mangrove swamps and coastal islands that provide a limited barrier against cyclones blowing in from the Bay of Bengal. The result will be greatly intensified storm surges, inflicting still greater damage on the country and displacing still more people.[104]

In 2010, the US director of national intelligence Admiral Dennis Blair reported to Congress that "for India, our research indicates that the practical effects of climate change will be manageable by New Delhi through 2030. Beyond 2030, India's ability to cope will be reduced by declining agricultural productivity, decreasing water supplies, and increasing pressure from cross-border migration into the country."[105]

The year Director Blair predicted, 2030, is now only 10 years away. To repeat: short of nuclear war, nothing that the great powers can do to each other remotely compares to what climate change threatens to do to us all even in the medium term.

2

A Perfect Storm: Climate Change, Migration, Automation

The fruits of successful nationhood are what attract migrants.[1]

—Paul Collier

The free institutions which sustain the life of a free and united people, sustain also the hatreds of a divided people.

—Lord Salisbury

UNTIL GLOBAL TEMPERATURES RISE to a point that the direct effects of climate change start to become catastrophic, the single most important threat posed by climate change to the security of the Western states and Russia is likely to be an indirect one: further increases in migration, with consequent increases in political radicalization, polarization, and state paralysis in Western democracies.[2] As noted in Chapter 1, in South Asia, where states are much more vulnerable to the direct effects of climate change, large-scale migration will come much sooner.[3]

The number of additional potential migrants as a result of climate change is impossible to calculate, because in most cases climate change alone is unlikely to be responsible for their displacement. The causes of mass migration include climate change and environmental factors together with poverty, oppression, over-population, and conflict. It does not make sense, therefore, to talk of "climate refugees" as such, or to create a new legal category for them.[4]

Rather, climate change will feed into other factors of environmental degradation and social tension, producing more conflicts like the Syrian civil war. In the words of the NATO Quadrennial Review of 2010, "While climate change alone does not cause conflict, it may act as an accelerant of instability or conflict, placing a burden to respond on civilian institutions and militaries around the world."[5] By increasing populist chauvinism (including climate change denial) and political divisions, migration is also making concerted action against climate change by Western states more difficult.

Recent years have seen a steep decline in support for European mainstream parties and a steep rise in Green and socialist parties, on the one hand, and populist nationalists, on the other. Increased migration will also worsen irretrievably the already existing divisions among present EU and NATO members. Should these trends continue, Europe will fall into deep crisis. On both migration and climate change, as in the United States, policy will swing wildly at each change of elected government. At best, countries will face the same kind of paralysis of government that has afflicted the United States. At worst, democracy will become unworkable and collapse.

As the miserable example of abortive democracy in the Middle East demonstrates, while democracy is very good at resolving secondary policy issues through the minority's acceptance of the legitimacy of majority decisions, it cannot resolve irreconcilable differences between large sections of the population concerning the basic cultural identity of the state and society.[6]

It is absolutely necessary, therefore, that on both sides politicians approach these issues in a spirit of civic nationalism, pragmatism, and reasonable compromise. If the Greens and the left continue their blind ideological commitment to open borders, they will feed white chauvinism, tear their societies apart, and make effective action to limit climate change impossible. If right-wing parties allow aversion to left-wing ideologies to blind them to the existential dangers of climate change, they will betray the vital national interests of their countries and doom them to eventual extinction. As far as US geopolitical interests are concerned, the decay of democracy in Europe and the United States is beginning to undermine the ideological and "soft power" basis of American global leadership.[7] In the medium term, therefore, the threat to the West is not of a catastrophe as a direct result of climate change but rather that climate change and migration will combine with other deep problems to produce a "perfect storm" of social, economic, and political crisis.[8]

The Marxistoid writers Michael Hardt and Antonio Negri have argued that a borderless world in which nation states are dissolved by universal capitalist globalization and movements of people will lead to the growth of a united international revolutionary movement of the poor (though Marx would have been more likely to call them the lumpen proletariat). As far as the possibility of creating an actual global movement along these lines is concerned, this is simply silly. Nothing of the sort is occurring.

Nor is it likely to occur, except in one scenario, which may be moving from the realm of science fiction to that of reality: namely, a future in which the rich genetically engineer themselves into what is in effect a new humanoid species, against which the mass of *Homines sapientes* then revolt.[9] Such a conflict might well help limit climate change in that a majority of CO_2-emitting humanity would be wiped out in the process. However, speaking as a *Homo sapiens* myself, this is a future for my human children and grandchildren that I would prefer if possible to avoid.

Hardt, Negri, Alain Badiou, and their like have not really thought about any concrete future scenarios at all. Rather, they are simply indulging in the old revolutionary fantasy: tear the existing order to pieces and something better is bound to take its place. After the experience of the past century, it should hardly need pointing out that this is a fantasy that has never turned out well. At least, though, writers like Hardt and Negri have the logical consistency of their own ideological lunacy.[10] It is much stranger to find supporters of the destruction of the nation state and national identities among writers—whether liberal capitalists like Parag Khanna or socialists like Naomi Klein—whose positions and hopes in fact depend utterly on the continued existence and effectiveness of such states.[11]

The Incoherence of the Debate on Climate Change and Migration

The debates in the United States and Europe on climate change, migration, and artificial intelligence are symptoms in themselves of the intellectual and ideological fragmentation of the West. Discussions of these issues have taken place very largely in isolation from each other, though the connections between them should be all too clear.

Thus, classical economists have made assumptions about the economic consequences of both climate change and migration that are often almost surreal in their indifference to social and political factors and the work of experts in these fields.[12] A majority of climate change activists, at least among liberals and socialists, have tried to downplay the relationship between climate change and migration, ignore the relationship between migration and the future of employment, and refuse to discuss the relationship between population decline in Western countries and a "de-growth" which they otherwise favor.

The desire of environmentalists simultaneously to issue dire warnings that climate change will increase migration and to downplay the results of migration can produce some curious effects. Thus chapter 14 of Christian Parenti's otherwise excellent book *Tropic of Chaos* is called "Golgotha Mexicana" and includes both a dire warning of environmental degradation and the threatened impact of climate change in Central America, and a truly terrifying (but apparently wholly accurate) portrait of drug wars, predatory police, corruption, rape, and murder in Ciudad Juarez, on the other side of the Rio Grande from Texas. The next chapter is titled "American Walls and Demagogues," which then bizarrely dismisses American fears about illegal mass migration from Mexico and the import of Mexican social, political, and criminal patterns to the United States as "paranoia" and "hysteria."[13]

Liberal environmentalists have taken this line even though the threat of increased migration due to climate change is clear in its general outline if not in its details or extent; and it also represents an important chance of appealing for electoral support in the West.[14] More than anything else, as opinion surveys clearly demonstrate, it is the indifference of leftist and center-right parties to the worries of their constituents concerning migration that has caused progressive parties to lose votes to the populist right and has contributed to a whole string of political disasters in Europe and the United States, including Brexit and the election of Trump.[15]

The only essay on migration in the standard handbook on climate change and society brushes aside concerns about climate change and migration as due to "the fear and cartographic anxieties of the affluent and the privileged that their 'orderly' spaces are going to be invaded *en masse*"; this claim is made in the face of increasing predictions that, on present trajectories, the number of international migrants (i.e., not including internally displaced people) may reach 400 million by 2050, due in part to climate change, though of course such estimates about the future are in the nature of intelligent guesses.[16] It may be remembered that the wave of Syrian refugees to Europe, which drove widespread populist radicalization and political destabilization in 2015, numbered only about one million.

It is true, of course, that the increasingly insecure and unemployed indigenous working classes of postindustrial Europe and America are still a great deal better off than the poor of Africa or South Asia; but for professors from elite Western universities to describe them as "affluent" and "privileged" is morally stupid, factually grotesque, and—should any of them read it—calculated to drive them into the arms of Donald Trump and the Front Nationale.[17]

As President Barack Obama warned the Democratic Party proponents of increased discrimination in favor of minorities,

> Most working- and middle-class white Americans don't feel that they have been particularly privileged by their race. . . . They are anxious about their futures, and feel their dreams slipping away. . . . To wish away the resentments of white Americans, to label them as misguided or even racist, without recognizing that they are rooted in legitimate concerns—this too widens the racial divide, and blocks the path to understanding.[18]

By the same token, populist nationalists are lining up behind climate change denial on ideological lines, even though both fear of migration and cultural conservatism ought to make them of all people committed to defend the global ecosystem. As for the separate public and academic debates on artificial intelligence, migration, and rebuilding industry for "green growth," very few of those involved in each debate pay any attention at all to the others, though the connection between them could hardly be more obvious.

The reason for this failure of cohesion and responsibility is essentially the same as the reason for the flight of Western security establishments back to the Cold War: a desperate fear on all sides of stepping outside comfort zones—ideological, cultural, social, political, and institutional—and confronting uncomfortable questions.[19]

As Paul Collier has written, opposition to immigration among the less skilled (or rather very often the highly skilled in a fading occupation) is often based on rational economic calculations and not on blind "xenophobia."[20] Yet among most liberals, social democrats, and Greens, the response has essentially been to tell actual and potential voters for the populist parties even more loudly what illiterate chauvinist idiots they are. Then they wonder why they are losing votes among the former working classes.

Climate Change and the Pressure to Migrate

Thomas Homer-Dixon a generation ago drew attention to the prospect that climate change and ecological scarcity would exacerbate conflicts.[21] As of 2018, however, there is little evidence of this leading to conflict between states. Instead, as Homer-Dixon predicted, environmental factors often have indirect effects in worsening existing internal tensions and conflicts, not only in the countryside but in cities where rural migrants clash along ethnic and religious lines.[22] As Western intelligence agencies predict, this in turn will increase already very high levels of migration.[23]

Thus, the Syrian civil war was due to multiple factors, including the failure to spread a sense of unitary nationalism in the Syrian population, anger at the tyranny of the Ba'ath regime and the domination of that regime by the Alawite Shia minority, and support for different factions by outside powers. In Syria and all over the Middle East, the revolts of the "Arab Spring" were also due to economic suffering and discontent, especially among impoverished recent migrants to the cities from the even more impoverished countryside.

We know that Syria experienced a severe drought in the years before the civil war and that bread prices across the Middle East rose steeply as a result of a worldwide rise due to droughts in grain-growing countries—and in particular China's purchase of huge amounts of wheat on world markets. It would be irrational and unhistorical not to assume that this had an effect in worsening the public discontent that led to the Arab Spring.[24] It is both wrong and unprovable to say that climate change alone "caused" that drought; but what we can say is that every single scientific prediction concerning the effects of climate change has suggested that among them will be the worsening of droughts and desertification in various parts of the world, including the Middle East.[25]

The same can be said of a range of conflicts in Africa.[26] Of course, climate change did not cause the old history of clashes between the Somali clans; or the Fur and Arabized tribes in Darfur; or the Turkana and Pokot in Kenya; or between Fulani pastoralists and Bachama, Tiv, and other farmers in central Nigeria (which by the middle of 2018 had claimed more than six times as many lives as the better known conflict with Boko Haram in the north and had displaced around 300,000 people).[27]

Nor has climate change been responsible for the steep rise in population that has put increasing pressure on scarce resources; nor the state weakness and economic backwardness that have made it impossible to create efforts at water conservation; nor the spread of automatic weapons that have made long-standing conflicts so much more deadly. Unlike in Australia, where changes in weather patterns are new and can with reasonable certainty be attributed to climate change, in West Africa (though not East Africa), desertification has been occurring for several decades.

What we can say with certainty is that in these semi-arid regions, access to water has always been critical to social stability and a source of many conflicts, both individual and collective. As in the Middle East, we can show that the great majority of the expert analyses concerning climate change predict worsened droughts in many areas.

It is also a fact that the population of sub-Saharan Africa has increased enormously over the past two generations and is expected to go on increasing very rapidly for decades to come, placing a greater and greater strain on resources, including water and fertile land. Between 1955 and 2018, the population of Nigeria rose almost five-fold, from 41 to 199 million. By 2050 it is expected to exceed 400 million, with a population density higher than England's, in one of the poorest large countries on earth. During the same period, the population of the African continent as a whole is predicted to rise from 1.28 billion today to 2.5 billion.[28] A combination of the rise in population with movement to cities situated in coastal areas means that by 2060, according to some estimates, between 26 and 36 million Africans will live in the floodplains of urban areas (up from a mere two million in 2000).[29]

Large parts of Nigeria are already suffering acute environmental degradation and are seriously threatened by climate change. The World Bank estimates that 30 to 70 million people in sub-Saharan Africa will be displaced in part due to climate change by 2050. On present trajectories for the rise in global temperatures, it expects numbers after that to rise more steeply. These people would initially of course be displaced within their countries, but it would be quite illogical, given patterns to date, not to assume that this will also lead to hugely increased numbers of migrants trying to reach Europe.[30]

West Africa is already one of the chief sources of illegal migration to Europe across the Mediterranean, which is causing not only internal political crisis within European states but leading to increasing friction between them—including between two of the original and key EU states, France and Italy. This threatens the survival of the EU in anything like its present form.[31]

Migration and Climate Change in South Asia

Migration within South Asia has already generated considerable violence in certain areas, which climate change can only worsen. Critiques of opposition to migration have focused on the West, but as Myron Weiner noted back in 1995, "An unwillingness to incorporate migrants into society is a feature of many Third World countries."[32] Thus the migration of millions of people from Zimbabwe and Mozambique to South Africa has led to repeated attacks on them by local people. On the other hand, all the countries in South Asia have benefited enormously from remittances sent home by migrant laborers in the Persian Gulf and emigrants to Europe and North America.

Migration for work or trade has a very old history in South Asia, with particular ethnicities and castes playing a leading role. Since the partition of the British Indian Empire, this migration has also acquired a trans-national dimension. Within both India and Pakistan, the movement of people has generated ethnic tensions, which also embody elements of class difference and urban-rural divisions. In South Asia as in Southeast Asia, this gave rise to demands that "sons of the soil" should be given precedence over migrants in state jobs and places in higher education.[33]

Thus, when Bombay was made the capital of the new Indian state of Maharashtra, ethnic Marathis "sought to turn Bombay into a Marathi city" and demanded that as the "indigenous" majority in Maharashtra, they should take precedence over the Gujaratis, Muslims, Christians, and Parsis who had dominated the city under British rule. These demands were given great additional force by the mass migration of Marathis from the countryside into the city. "Other ethnic communities could remain in the city and even retain their own distinctive cultural identities, but it should be understood, argued the Marathi nativists, that the city belonged to Maharashtra, that native Marathi speakers had special rights within it, and that all who resided in the city had to speak Marathi at least as a second language."[34]

More recently, this Marathi ethnic sentiment has flowed into the virulently anti-Muslim Shiv Sena Party, an ally of Narendra Modi's Bharatiya Janata Party (BJP), and contributed to violent communal clashes and pogroms in the city. As Myron Weiner notes, some of this at least could have been avoided if Bombay had remained a territory under central government control.[35]

The consequences of migration and the loss of federal territory status have been far worse in Pakistan's largest city of Karachi, which was the capital of Pakistan until 1967. Karachi is situated in the province of Sindh, but under British rule it became very different from its Sindhi Muslim hinterland ("interior Sindh" in the rather revealing phrase of Karachiites) and was dominated by Sindhi-speaking Hindus.

When these were driven out at Partition, they were replaced by Muslim migrants from India, the "Mohajirs." With higher standards of education and a much stronger commercial tradition, the Mohajirs completely eclipsed the indigenous Sindhis in Karachi, leading to bitter resentments among the latter. When the capital moved to Islamabad, and in 1971 a Sindhi-led government took power in Pakistan, the Sindhis asserted themselves strongly against the Mohajirs.

This led to repeated ethnic clashes, which among other things caused a violent polarization of the students at Karachi University from which that institution has never recovered. In the meantime, migration by ethnic Pashtuns from Pakistan's impoverished northwest also increased. This was given a great boost by the Afghan war of the 1980s, which led to some three million Afghan refugees fleeing into Pakistan. The Afghan war also caused an immense spread of automatic weapons within Pakistan.

In the mid-1980s in Karachi, this toxic mixture led to the creation of the Mohajir Qaumi Movement (MQM) dedicated to consolidating Mohajir domination of Karachi by a combination of modern political organization and paramilitary force. Since then, the city has been periodically racked by episodes of ethno-political violence, interspersed with equally ferocious crackdowns by the central state and the military. After the US invasion of Afghanistan in 2001, a new violent factor was introduced in the form of Islamist terrorism, stemming chiefly from within the Pashtun population.

At the time of writing, a tenuous peace has been established by military repression. The MQM has been broken up (thereby reducing ethnic violence but also the efficiency of municipal government). Despite bitter anti-Mohajir rhetoric, the Sindhi elites have long since in effect been bought off by being allowed to milk the economy of Karachi through their domination of the Sindh provincial government. Ever since I first visited the city in the 1980s, I have had a baffled affection for Karachi and a deep sense of frustration and tragedy that a metropolis whose economic and cultural dynamism could have helped to lift Pakistan as a whole out of poverty has instead been paralyzed as a result of ethnic strife fueled by migration.[36]

The other area of South Asia that has experienced very large-scale violence as a result of migration is northeast India and the Arakan region of Myanmar. This has been largely due to reactions against Bengali migration, initially from the British province of Bengal and more recently from heavily over-populated Bangladesh (considered the major country of the world most at risk from climate change).[37]

Since the 1970s, the region has seen periodic episodes of violence as local populations react against Bangladeshi migration. They fear the fate of the indigenous population of the Indian state of Tripura, where long-standing Bengali migration (especially the flood of Hindu refugees after 1947 and during the Bangladesh war of independence of 1971) has reduced the indigenous population to barely a quarter of the total.

In neighboring Myanmar, the same fear among ethnic Burmese (assiduously stoked by the military) contributed to the savage campaign against the Bengali-speaking Muslim Rohingya minority starting in 2017, which led to more than 700,000 fleeing into Bangladesh. Ever since Burmese independence in 1947, the alleged migrant origins of the Rohingya have been advanced as grounds to deny them full Burmese citizenship.

The Rohingya presence in the province of Arakan dates back to the 15th century. Their numbers grew considerably, however, as a result of labor migration under the British Empire in the 19th and early 20th centuries (as in Sri Lanka, Malaysia, Fiji, and elsewhere).[38] With the exception of Bangladesh, Myanmar's other neighbors have generally refused to accept Rohingya refugees. Since these neighboring countries have not ratified the UN Convention on Refugees, they are under no legal obligation to accept these migrants.

India has been especially unsympathetic, as part of its general policy of attempting to stop Bangladeshi migration—an effort that has been further strengthened in recent years by Hindu nationalist fears of growing Muslim numbers in India. The National Registry of Citizens (NRC) has been introduced in part as a way of identifying and possibly deporting them. Amit Shah, president of the Hindu nationalist Bharatiya Janata Party (BJP), a close ally of Narendra Modi and a leading fomenter of Hindu chauvinism, said of the Bangladeshi migrants during the Indian election campaign of 2019, "These infiltrators are eating away at our country like termites. The NRC is our means of removing them."[39]

Since the 1990s, India has constructed a barbed-wire barrier along the border with Bangladesh, backed up with what has been alleged to be a "shoot to kill" policy by the Indian Border Security Force. Between 2001 and 2018, this led to 1,109 deaths of Bangladeshi nationals, though deaths have gone down recently due to better cooperation between the Indian and Bangladeshi governments.[40]

A former head of India's intelligence service (the Research and Analysis Wing, or RAW), Sanjeev Tripathi, was however unapologetic:

> Effective policing of such a long and complex border is difficult. India has sought to establish greater control over the border by erecting barbed-wire fencing. The plan to extend such fencing along the entire border should be implemented at the earliest possible time. A border fence may not be a fool-proof method of preventing infiltration, but there is no better first step.[41]

In 1991, a cyclone, which drove a six-meter storm surge, killed at least 138,000 Bangladeshis and left 10 million homeless. During Cyclone Siri in 2007, better early warning and evacuation meant that only around 15,000 died, but millions were displaced and a large part of Bangladesh's harvest was destroyed, leaving the country dependent on international food aid. Given this vulnerability, it is not hard to see how even a relatively small rise in sea levels could drastically worsen the impact of such storms, as well as changing—perhaps radically—the present course of rivers.[42] Given the degree of tension and violence already surrounding Bangladeshi migration, any major increase in that migration is bound to fuel further conflict. We know the violence that a few million migrants (or the fear of them) has caused in South Asia. The violence caused by tens of millions might well be on a genocidal scale. It would dwarf the casualties in all the conflicts between India and Pakistan put together. In fact, it would probably dwarf any imaginable conflict in South Asia short of all-out nuclear war. It would also create immense new pressure to migrate to the West.

Migration to Russia

Increased migration is also the greatest long-term threat to Russia from climate change and the chief answer to those Russians who think that their country will benefit from global warming while avoiding its negative consequences (apart from the fact that the idea that Russia could prosper amid a world economy wrecked by climate change is obviously absurd). As a huge Eurasian country with the longest land borders in the world. Russia would not be able to insulate itself against the consequences of the collapse of states to its south.

Russia's experience and dilemmas concerning migration and integration echo to some extent the experience of the rest of Europe. In the generation after the collapse of the USSR, Russia depended heavily on migrants from the other former Soviet republics to replenish its own falling population. However, in the 1990s, these migrants were largely ethnic Russians or Russian speakers. As a result of this move of members of the Russian diaspora to Russia and the pull of Russia's stronger economy, at the start of the 21st century some 8.3 percent of the Russian population were migrants, the highest proportion in the world after the United States.

The question now is whether Russia should accept increasing numbers of Muslim migrants, many from the former Soviet territories in Central Asia, who from 2010 on made up about half of immigrants to Russia.[43] Recent years have seen frequent demonstrations and episodes of violence against Muslim immigrants. As in Western Europe, anti-immigrant feeling has been a leading theme of Russia's radical nationalist parties. The government has responded with tougher measures to regulate migration and deport migrants found guilty of even minor offenses.[44]

The fear that Russians are being increasingly outnumbered by Muslims from Central Asia and the Caucasus dates back to the 1970s, when the consequences of falling birth rates in Russia, Ukraine, and Belarus and falling death rates in Central Asia first became apparent. Public expression of these fears was restricted under Soviet rule but played a certain part in the willingness of Russians to see the USSR disintegrate.

Since the end of the Soviet Union, hostility to Muslim immigration has become a significant factor in Russian politics and in the debate on Russian national identity. The section of the Russian opposition led by Alexei Navalny has on occasions tried to use this as a tool in their struggle against President Vladimir Putin.[45] It was even suggested that Russia should withdraw from the northeast Caucasus, both to extricate itself from the Chechen conflicts and to allow the expulsion of Chechens and Daghestanis from Russia. Public fear of Muslim immigration has been increased by Islamist and Chechen separatist terrorism.[46] According to a Levada opinion poll, between 2002 and 2013 the proportion of respondents declaring that "Russia is for ethnic Russians" grew from 55 to 66 percent.[47]

Putin himself, in line both with his Soviet heritage and his desire to unite republics of the former USSR in a Eurasian Union under Russian hegemony, has opposed Russian ethnic chauvinism and been more attached to the idea of a multi-ethnic and multi-religious Russian identity. His combination of this with an insistence on loyalty to the state and the leading role of Russian culture has some potentially useful lessons for the West:

> Historically, Russia has been neither a mono-ethnic state nor a US-style "melting pot," where most people are, in some way, migrants. Russia developed over centuries as a multinational state, in which different ethnic groups have had to mingle, interact and connect with each other. . . . We need a national policy strategy based on civic patriotism. There is no need for anyone living in Russia to forget their religion or ethnicity. But they should identify themselves primarily as citizens of Russia and take pride in that. No one has the right to put their ethnic or religious interests above the laws of the land. At the same time, national laws must take into account the specific characteristics of different ethnic and religious groups.[48]

Putin's stance is also due to his desire to strengthen Russia's great-power status by strengthening the Eurasian Union.[49] This alliance requires the membership of Central Asian Muslim states, and (as with East European desire for EU membership) the desire of the Tajiks and Kirghiz to be part of the union is very largely due to their desire for free labor movement to the vastly wealthier Russian Federation.

Some of the strongest hostility to Muslim immigration has come from ethnic minorities in Russia and in particular the "titular nationalities" of Russian autonomous republics like Yakutia (Sakha) and Kalmykia.[50] These people fear that their already limited numbers are in even greater danger of being swamped by migration. They also fear that—as in Western Europe—rising Muslim numbers will push the ideology of the Russian state away from multi-ethnicity and toward assimilation.[51] This would pose a threat to the ethnic identities of all the Russian indigenous minorities, tenuously preserved through centuries of Russian and Soviet rule.[52] The potential contribution of climate change to Muslim migration and these growing tensions is obvious, given concerns about future drought in Central Asia.

The deputy governor of the Siberian region of Khanti-Mansiysk once told me that he welcomed climate change because it would allow oranges to be grown in Siberia. I pointed out to him that the Siberians would probably have to share those oranges with several tens of millions of newly arrived Uzbeks, Afghans, Pakistanis, and possibly Chinese. His mouth opened and stayed that way for a while. "Oh," he said. "I didn't think of that."

Migration, Artificial Intelligence, and the Green New Deal

When it comes to migration to the West, the danger is not of climate change creating a crisis. The crisis concerning migration and the reactions against it are already all too manifestly with us. The questions are how much worse it will become as a result of climate change, and how much damage will result for efforts successfully to limit climate change. With regard to climate change there is a double and reciprocal relationship: climate change is bound to increase migration, though by how much we do not know; and migration is already helping to undermine the political unity of Western states that is essential if these states are to adopt policies against climate change that will involve sacrifices by their populations.

Especially since 2008, there has been a growing belief, in Dani Rodrik's words, that "democracy and national determination should trump hyper-globalisation. Democracies have the right to protect their social arrangements, and when this right clashes with the requirements of the global economy, it is the latter that should give way."[53] This applies as much to migration as it does to finance and trade.

So any comprehensive thinking about limiting carbon emissions, promoting green growth, and restoring social solidarity has also to deal with the issue of migration. The Green New Deal idea is an essential step in winning over the working classes to support action on climate change, and European variants need to be urgently developed.[54]

However, when they promise to create numerous well-paid and secure jobs, such programs are especially incompatible with the advocacy of open immigration by many of the people advocating a Green New Deal.[55]

One of the most worrying things about continued mass migration to the West is that it is likely to coincide with greatly intensified automation of the economy, including the development of some kind of artificial intelligence (AI). The resulting job losses may be on a colossal scale, and if present patterns continue, these jobs will be replaced by worse-paid and more insecure jobs, or not replaced at all. Any serious long-term strategy for a Green New Deal will need to address all of these issues together.

If even the medium-range predictions for automation are correct, then they radically contradict the economic argument for mass migration of unskilled workers, endlessly touted by migration advocates: that it is necessary to provide new workers to compensate for falling birth rates in the developed world. Trying to compete with artificial intelligence through ever cheaper migrant labor is the worst idea of all.[56] Japan by contrast has always preferred artificial intelligence over the import of migrant labor, which has been minimal (which by the way also offers a counter-argument to the idea that "migration cannot be stopped"),

It is true that in recent years the Japanese economy has stagnated—but so has that of Europe, so a lack of migrants to Japan cannot be the explanation for this. Furthermore, many environmentalists are hostile to the obsession with economic growth and have warned that if our economies are to be sustainable in the long run, we must move from growth to stability or even de-growth. If so, it is hard to see why Japan should be urged to accept migrants for the sake of further growth rather than managing population decline and economic stability in a very rich and over-populated country.

The Japanese population ceased to gow in the 1980s, and since 2011 has begun a steep decline. This demographic trend has been accompanied by very low economic growth. Nevertheless, there is very little support indeed in Japan for a relaxation of Japan's severe restrictions on immigration for the rest of Asia. It would seem that by an informal and perhaps only semi-conscious consensus, the Japanese have decided that "de-growth" is in fact an acceptable price to pay for the maintenance of their ancient ethnic homoegeneity.

One response that has been suggested is some version of a universal basic income (UBI).[57] Such a program could be part of the Green New Deal. It would be entirely compatible with action against climate change, because such redistribution and agreed, state-backed limited incomes would include a strong spirit of austerity and collective sacrifice. It would require a state-imposed program of heavy taxation and redistribution on a scale not seen since wartime.[58]

Like the Green New Deal, such a program is, however, quite incompatible with continued high levels of migration. In the first place, it could only be implemented on a national scale. The idea of wealthier nations distributing their wealth in this way to recently arrived humanity in general is simply fanciful. As a national program, it would create the strongest possible incentive for existing citizens (doubtless including many former immigrants) to refuse entry to migrants, or, if they had to be admitted, to make it extremely difficult for them ever to gain citizenship.

For many years, fears about added strains on social welfare, health, housing, and school systems have been among the chief causes of opposition to migration.[59] Despite some concrete estimates (like that by the German economics ministry that the cost of integrating the Syrian and other refugees would amount to more than 400 billion Euros), the evidence for these costs has always been somewhat anecdotal and difficult to pin down.[60] Under a UBI system, by contrast, anyone with a pocket calculator could work out how much his or her basic income would drop for every given new percentage of migrants. Moreover, since these payments would be shares in a national economy accumulated over many generations, this would raise still further the ethical question of why new citizens should be given such shares.

I have become vividly aware of these issues as a result of my residence in Qatar, which has the highest proportion of migrant non-citizens to indigenous citizens of any country in the world (over 85 percent), followed by the other oil-rich Gulf states. Qatar also has the most restrictive laws concerning the granting of citizenship of any country in the world, although with barely 300,000 people and 12 percent of the population, the indigenous population is in great need of new blood, new skills, and a new work ethic.

The citizen body, however, is deeply opposed to relaxing the naturalization laws (even for other Arab residents), because Qatar's gas wealth has allowed it to create what amounts to a system of UBI for its citizens—not by that name, but in the form of guaranteed state jobs, interest-free mortgages and loans, and so on.[61] Citizens know very well that these benefits—on which many have come to rely completely for their living standards—would be severely diluted by any significant expansion of the citizen body.[62]

If the advocates of migration have ignored the automation issue altogether, the mainstream economists have responded with what can only be called their own version of magical thinking, based on startlingly flawed and superficial historical parallels. The argument has been that previous economic revolutions produced masses of new jobs in wholly new industries and services to replace those lost in agriculture and traditional industries, so the artificial intelligence and computerization revolution will do the same. The point about the gathering economic revolution, however, is that there is no logical reason whatsoever why the new industries and services will not themselves also be automated. On the contrary, precisely because they are new, they are even more likely to be automated than the old ones.

According to Carl Benedikt Frey and Michael A. Osborne:

> The reason why human labour has prevailed [in the past] relates to its ability to adopt and acquire new skills by means of education. Yet as computerisation enters more cognitive domains this will become increasingly challenging. . . . High-skilled workers have moved down the occupational ladder, taking on jobs traditionally performed by low-skilled workers, pushing low-skilled workers even further down the occupational ladder and, to some extent, even out of the labour force.[63]

These authors estimate that 47 percent of total US employment is in the high-risk category, meaning that associated occupations are potentially automatable within a decade or two; and four years after their work appeared, one can see that this is indeed happening.[64] A 2015 study estimated that a large majority of the jobs lost in US manufacturing industries between 2006 and 2013 were lost due to artificial intelligence. In August 2019, Western banks began mass reductions of their staff, in part due to the impact of automated trading.[65] In other words, artificial intelligence is already one cause (though, of course, not the only one) of the stagnation or decline of middle-class incomes and the increasing numbers of US men who have fallen out of the workforce (by some estimates, four times the official unemployment figure).[66]

As for unskilled or semi-skilled migrants, the loss of secure, well-paid jobs in the economy condemns most of them permanently to poor, insecure jobs. These in turn trap migrants and their descendants in impoverished ghettoes, from which bad schools, crime, low property prices, and social isolation make it very difficult to escape. There is simply no excuse for ignoring the reality of this picture, given what we see in the banlieues of France (where youth unemployment already runs at more than 40 percent), the former mill towns of northern England, and the former steel towns of Belgium and northern France, where unemployment in the indigenous population is running at over 20 percent.[67] And of course, migrants are in competition with increasingly poor members of the indigenous working-class population for the same low-paid and insecure jobs.[68]

In the United States, free-market economists have long talked quite openly of using immigration to drive down wages and discipline workers;[69] and while continuing immigration may have benefited the US economy as a whole, the inescapable fact about US economic growth since the 1970s is precisely that it has *not* benefited unskilled and semi-skilled American workers. This is in part because of the constant downward pressure on wages due to immigration, which in turn has been partially enabled by the use of migrant labor by employers to weaken the trade unions.

Even the more intelligent of the optimistic predictions for job creation due to artificial intelligence admit that considerable state intervention will be needed to smooth the transition and support the losers from the process, as well as to educate the workforce in new skills.[70] Nobody has come up with a convincing (at least to me) account of what is going to happen to the unskilled (or technologically unskilled), let alone why the European economy needs more of them.[71]

The impact of artificial intelligence and computerization also raises the prospect that huge populations in India and elsewhere, which have been looking forward to repeating the economic success of the East Asian "tigers" through work in cheap manufacturing and services, will find these jobs taken by machines (as is already happening in that arch-field of Indian economic success, telemarketing).[72] If so, even newly "successful" Asian economies will still contain vast impoverished populations. Not only will this pose serious threats to stability and state effectiveness in parts of Asia, but it will presumably mean that many still aim to migrate to Europe—unless of course Europe becomes so impoverished and violent in turn that it is no longer a desirable destination.

Not surprisingly, the failure of both the Democrats and the Republican elites to address the twin issues of migration and the decline of the working classes eventually produced the disastrous swing to Donald Trump (incidentally, it is very suggestive that whereas in the past the overwhelming majority of Americans defined themselves as "middle class," by 2018 almost half were describing themselves as "working class"—which is why I have used that term for these sections of the American population in this book).[73]

Artificial intelligence threatens in effect to turn a large part of present middle-class employment as well into unskilled, insecure, part-time labor compared to the skilled permanent labor that can be done by computers. This will create large numbers of the most dangerous political actor in modern history—the unemployed graduate.[74]

Migration, Climate Change, and State Paralysis

If climate change helps drive migration, migration also helps hinder serious action against climate change in general, and a Green New Deal in particular. On the one hand, as this book argues, if populations are to make sacrifices in the struggle to limit climate change, they need to be assured that the state will support their essential needs and that society as a whole will be involved in the struggle. Hence the need for a Green New Deal (quite apart from the need for it on wider social, moral, and political grounds) involving a great deal of redistribution of incomes.

But there is now an abundant body of statistical evidence demonstrating a relationship between the growing diversity of populations and a decline in social trust, social solidarity, and the willingness of majority populations to pay taxes to support social welfare for minorities and especially recent arrivals,[75] showing that, in other words, the welfare state has to be a reasonably homogenous state.[76] Thus, in Britain, willingness to pay higher taxes to support social welfare declined from 58 percent of those surveyed in 1991 to only 28 percent in 2012.[77]

Robert Putnam has researched this development in US communities: "Diversity seems to trigger not in-group/out-group division, but anomie or social isolation. In colloquial language, people living in ethnically diverse settings appear to 'hunker down'—that is, to pull in like a turtle."[78] Strikingly, Putnam has also shown that an increase in numbers diminishes trust among the immigrants themselves—one reason presumably being that as Pakistanis, Bangladeshis, Mexicans, and others form big, largely closed immigrant communities, those communities reproduce the low levels of interpersonal trust and high levels of familialism (which tends to lead to clientelistic corruption) that have helped to cripple social harmony and economic development in their countries of origin.

The historic nature of the United States as a country of high (and ethnically new) immigration and acute racial division has been advanced as a key reason for why the nation has never developed a welfare system on the scale of European countries.[79] It can also be argued that the only time this was successfully advanced, the decades after the New Deal (followed by Eisenhower and Nixon as well as successive Democratic administrations) coincided with the restrictions on immigration between 1924 and 1964, which allowed the stabilization of an ethnically diverse but culturally homogenous white working class, thereby creating the decades of secure black industrial employment and rising working-class living standards that came to a halt in the 1970s.

As Michael Lind warned as early as 1995 in *The Next American Nation*, it will only be possible to create a New Deal coalition based on a sense of national social solidarity across racial lines if existing workers are reassured that their wages and job security will not continuously be driven down by legal and illegal immigration.[80] This will be especially true if their jobs are also coming under intense pressure from artificial intelligence.

Migration also raises wider issues of national unity. As Chapter 5 of this book argues, a Green New Deal, like the original New Deal, will require sweeping electoral majorities to create it, and a solid national consensus to maintain it. More widely, strong elements of social solidarity will be needed both to legitimize sacrifice in the struggle against climate change and to sustain states and political systems in the face of the shocks that the next century will bring. It is quite obvious that in the United States, Britain, and Europe, the backlash against migration is pushing in exactly the opposite direction, toward deeper and more embittered polarization. Action against climate change has been one of the victims of this tragic development. Advocates of mass migration are doubtless well-meaning, but as Ian Bremmer notes, "It is not racist to acknowledge that the best of intentions sometimes produces terrible consequences."[81]

If further migration to the West can be kept within reasonable limits, and without sudden massive spikes like the Syrian refugee crisis and the overthrow of the Libyan state, then there is a good chance that migrants and their descendants can be successfully integrated. In *Whiteshift*, Eric Kaufman writes of a desirable future in which core national groups and national traditions will continue but will be open to immigrants and their descendants, above all through the oldest and deepest of all forms of integration, intermarriage.[82] The "whites" in the United States would get somewhat browner, as would the English, the Germans, the Danes, and so on—and no great tragedy in that. I agree therefore with David Miller that ultimate genetic origin is absolutely irrelevant. "All that matters is that the melding together of different 'races' should have produced a people with a distinct and common character of its own"—though not one that has to be monolithic, and to be shared, or equally shared, by everyone.[83]

To quote that great American statesman Senator Jay Bulworth, "All we need is a voluntary, free spirited, open-ended program of procreative racial deconstruction. Everybody just gotta keep xxxxin' everybody 'til they're all the same color."[84] Or in Daniel Defoe's version of the origins of the English: "Thus from a mixture of all kinds began / That het'rogeneous thing, an Englishman."[85] This, after all, was very much the pattern by which previous waves of European migration to the United States and between European countries were integrated into their new countries without radically changing those countries' core traditions.

It is true that the idea that "Europe has always been an area of migration" is arrant nonsense for the period between the Dark Ages (when people did indeed migrate—armed with spears) and the 19th century. However, the industrialization of Western Europe saw huge numbers of Poles moving to Germany, Irish to England, Italians to France, and so on, and their descendants fully assimilating.[86]

Race is an obstacle (especially in the United States) but by no means an insuperable one, as the great land empires of Eurasia demonstrate. Anyone who has spent any time at all in Russia or China will know from observation that people calling themselves "ethnic Russians" or "Han Chinese" are in fact drawn from a very great mixture of racial origins (I once attended a gathering of Russian Cossack leaders where the facial types stretched from Scandinavian to Assyrian). Yet they are also part of extremely strong national identities and traditions.

I am reasonably confident that in the long run this will also be the pattern for Latinos in the United States (after all, the Bush dynasty has Mexican members and a large section of the Republican leadership is of Cuban descent). It is true, though, that the Democrats' fetishization of identity politics is a considerable obstacle. As the proportion of Latinos in the population soars (from 6 percent in 1960 to 17 percent in 2018, with projections of over 25 percent by mid-century), designating them as a separate "race" is sheer madness, which plays into the hands of white nationalists arguing that whites will soon be turned into an oppressed minority by "other races" and core American traditions destroyed.[87] Latinos are mostly a mixture of European and American Indian and could just as well be called "white." They are also, of course, predominantly Christian.

In Western Europe, the assimilation of people of West Indian, African, and East Asian descent is going reasonably well, including through intermarriage (or at least cohabitation). Integration into the political elites is rather slow, but in the British Parliament elected in 2017, 52 Members of Parliament (MPs), or almost 9 percent, came from racial minorities. The government formed by Boris Johnson in 2019 was the most racially diverse in the history of Britain and Europe—and this, impressively, was a Conservative government. Social and cultural acceptance has been greatly helped by the worlds of sports, music, and entertainment. In Britain, black and mixed-race actors, singers, and TV stars like Idris Elba, Thandie Newton, Freema Agyeman, and Alesha Dixon are increasingly seen simply as "British." The London underclass also seems to be becoming multi-racial, or at least so it would seem from the London riots of 2011.

There are, however, a number of conditions necessary for this process to succeed. The first, as Kaufman emphasizes, is that core national traditions remain strong, so that migrants have something to assimilate into. American ideological nationalism has deep roots in American (and before that British) history; but it was also rigorously instilled through the school system from the later 19th century onward, precisely as a way of integrating immigrants.[88] In Europe, national identity was also systematically propagated through the school system and conscript armies.

A second key issue is numbers of migrants and the time period in which they arrive. Even very large numbers can probably be integrated successfully over a long period of time. This is necessary to give the indigenous population time to accustom itself to the process and to allow both the migrants and the indigenous population time to change culturally. It is also necessary if states are to carry out the economic, social, and ecological reforms necessary if economic hardship and climate change are not to combine with migration to destroy Western democracy.

By contrast, large numbers of migrants in a short period of time are much more likely to form cultural ghettoes, to marry within their own communities rather than with the majority population, to provoke panicked reactions in the indigenous population, and to divide political systems in a way that makes agreement on reform much more difficult. This is why a new surge in migration triggered by climate change (or rather, famines and conflicts exacerbated by climate change) would be so extremely dangerous to the political orders in Europe and the United States—as the political consequences of the Syrian refugee crisis have so clearly demonstrated.

A very large proportion of migrants to Europe will continue to come from the Muslim world, parts of which are also particularly riven by internal conflicts and particularly vulnerable to the effects of climate change. And with regard to Muslim populations in Europe, there is a set of issues that puts many of them in a quite different category from the Latinos in the United States.[89] There is the malign history of Christian-Muslim conflict, constantly evoked by both Islamist and white chauvinist extremists and terrorists. Conservative Islam, where it prevails, is an insuperable obstacle to the intermarriage and cultural assimilation advocated by Kaufman. The most that can be hoped for in this case is peaceful co-existence in essentially separate communities. As Innes Bowen writes in her study of British Muslim clerics, "Figures like Suleman Nagdi and Kaushar Tai [two conservative but non-extremist Muslim leaders in Leicester] may be fully engaged with the institutions of mainstream society, but their role is not to encourage integration. Rather it is to protect the ability of Muslims to live as a religious minority, fully practicing and expressing their faith."[90]

Intermarriage of Muslims in Europe with indigenous populations is a mixed picture, proceeding quite well in France (despite the deep poverty of many French Muslims), but proceeding only slowly, or very slowly elsewhere.[91]

The growth of separate communities is encouraged by the tendency of certain Muslim ethnic groups (like the Pakistani Mirpuris of Yorkshire and the Bangladeshis of London) to cluster together physically and intermarry with people from their places of origin. The existence of such communities, often speaking their own language and dependent for news and entertainment on the television of their countries of origins, further increases the social, economic, and geographical isolation of migrants and their descendants.

Trevor Philips, former head of the UK Commission for Racial Equality, warned in 2005, "We are sleepwalking our way to segregation. We are becoming strangers to each other, and we are leaving communities to be marooned outside the mainstream."[92] The growth of such very visibly culturally different communities increases the anxiety of indigenous populations.

Numbers are once again key. In Britain, where the census records religious affiliation, the proportion of Muslims in the population has gone up by over 50 percent each decade since the 1960s.[93] Even the medium projection of the Pew Research Center is for 16.7 percent of the British population to be Muslim by 2050, with 17.4 percent of the French population, 20.5 percent of the Swedish population, and 11.4 percent of the population of the European Union as a whole. If migration continues at a high rate (as it will if climate change drives famine and conflict in parts of the Muslim world), then Pew estimates that by mid-century, 30.6 percent of Swedes and almost one fifth of British, French, and Germans will be Muslim.[94]

This implies that certain areas of countries will have substantial Muslim majorities by mid-century. And migration, like climate change, will not stop in 2050. If these estimates are accurate and migration increases substantially due to climate change, by the end of the present century, Muslim pluralities in several countries seem possible, together with very large majorities in certain regions, like Yorkshire.

Such a development would be a matter of concern for many English people even if that minority were fully integrated and economically successful, which a large part of it most certainly is not and will not be if numbers continue to rise and automation has the effects predicted above. The role that a Muslim Yorkshire will play in a future British polity is something that few people so far have been willing to think about, but if things go on as they have been for the past two generations, in another generation this change in Yorkshire and other parts of Britain will have to be given consideration. As the British census of 2011 recorded that the white British population of London had dropped to only 44.9 percent—a full generation or more before anyone had predicted this—this is in no sense a fantastical prospect. And the result of mass migration in the generation before 2016 was the disaster of Brexit.

One cannot predict with certainty that the growth of large and solid Muslim territorial communities would cause a deep crisis for the British political system. What one can say is that there is no historical precedent for such a transformation under democracy and in peacetime; and where such demographic changes have taken place under imperial rule (the USSR in the Baltic States, the British Empire in Ceylon, Malaya, Fiji, and elsewhere), the results have varied from the difficult to the catastrophic.

There is a clear possibility that high levels of migration combined with cultural differences, economic change, and climate change will cause deeper and deeper divisions in Western societies, eventually leading to the demise of democracy. US politics over the past generation have been a graphic illustration of this danger: the political system will become largely paralyzed, because of an endless, struggle between irreconcilable camps—a kind of permanent political trench warfare—with great battles over great issues, but also constant debilitating skirmishes over symbolic ones; for example, in the United States of America, the Confederate flag, and in France the ban on headscarves.[95] This is becoming reminiscent of the endless squabbles over names and statues that characterized the Habsburg Empire and Yugoslavia in their last decades.

As Paul Collier has written,

> Lifestyles like that of my family [i.e., those of the educated and migratory cosmopolitan elites] are dependent, and potentially parasitic, on those whose identity remains rooted, thereby providing us with the viable societies among which we choose. In the countries on which I work—the multicultural societies of Africa—the consequences of weak national identity are apparent.[96]

I can most certainly say the same of Pakistan, Afghanistan, and large parts of the Middle East. In particular, Pakistan exemplifies the difficulty of creating a tax-paying ethic in the population without a sense of common identity and mutual responsibility. Since any conceivable program of response to climate change (whether prevention, mitigation, or adaptation) will require considerable increases in taxation, this should be of immense importance to environmentalists.

The relationship of migration to climate change is therefore a reciprocal one. If the rate of migration to Western societies can be slowed, then there is still a reasonable chance that integration may succeed and effective democratic government be preserved. For while I have drawn attention to the problems involved in the integration of Muslim minorities in Europe, there have also been some notable success stories, both economically and politically. A generation ago, it would have seemed hardly possible that in Britain the Labour Party mayor of

London and the Conservative chancellor should both be Muslims, and the Conservative home secretary an Indian Hindu by descent.

However, if increased migration combines with massive socioeconomic and cultural disruption due to artificial intelligence and unchecked climate change, then the result may be to return Europe to a previous, infinitely darker era. For in bandying about the term "fascist" to describe Donald Trump and the populist right in Europe, liberals and Greens risk obscuring the infinitely greater real evil of fascism. Nazi ideology had its roots in German Central Europe from the 1870s to the 1890s, when rapid economic, social, and cultural change (involving the destruction of many traditional occupations and their accompanying social status) combined with economic recession and mass migration.

The movement of Slavs to previously German-dominated cities transformed these cities linguistically and culturally. The migration of Jews from Eastern Europe and the Russian Empire created a dreadful new mixture of traditional anti-Semitism with modern economic resentments and political fears. It is true that to bring Nazism to power required the carnage of the First World War and a very deep economic depression; but as this book among many others has sought to warn, the future impact of climate change may be equivalent to that of war.

3

Nationalism and Progress

Conservation is a great moral issue, for it involves the patriotic duty of ensuring the safety and continuance of the nation.[1]

—Theodore Roosevelt

FOR A WESTERN STUDENT of nationalism to live outside the West over the past generation has been a rather schizophrenic experience. Until recently, the vast majority of Western intellectuals described nationalism not just as an artificial "construction" of modern elites, but also as an inherently bad thing—an attitude that had its roots in an understandable backlash against the impulses that produced the First and Second World Wars.[2] The nation state itself was widely denounced as an archaic relic of the past that would and should make way for international institutions and networks created by globalization.[3] Meanwhile, in the countries where I was working, responsible elites were bitterly regretting the weakness of national feeling and making assiduous attempts to strengthen it, fearing that the lack of it would lead at best to state paralysis, at worst to state failure.

Among my fellow Western journalists and analysts in think tanks, I could see how the belief in the wicked and the artificial nature of nationalism had become part of Western educated culture in general. In any given case the instinctive and automatic trope (often completely without regard to the evidence) was to describe the nationalism concerned as a result not of history and mass culture but of cynical manipulation by self-interested elites. This was also an immensely useful approach for Western policymakers, for any reaction by other nations against Western policies could be portrayed as the product not of genuine national pride and a sense of national interest but as a mass delusion created by wicked rulers.

This was also an old colonial trope (the corrupt "headman" or "chief") in new clothing. It also had the merit of making the analysis of other countries much less demanding, as a narrative of wicked rulers was conceptually so much easier than the alternatives. It also, unfortunately, meant that such analysis almost never accurately reflected any major changes, making it largely useless as a basis for policy.

Underlying these attitudes was a belief that nationalism is not only bad but restricted to backward peoples, and doomed therefore to melt away in the benevolent sunlight of Western-led globalization, whether of the capitalist or Marxian variety, or the curious mixture of both that characterizes much liberal thought today.[4] This no doubt explains the strength of the reaction against my book on American nationalism when it first appeared in 2004, not just on the part of the right (from whom I expected nothing else) but from many liberals as well. For if nationalism remained a critically important force in the country that supposedly epitomized globalized modernity, and if the liberal "internationalists" themselves in fact embodied their own version of messianic civilizational nationalism, then an entire liberal worldview and the personal, class, and national complacency it engendered would begin to tremble.

Nationalism and Legitimacy

Stemming as they do from the liberal internationalist or Marxian traditions, a majority of environmentalist writers and activists have shared in the hostility of these traditions to the nation state and nationalism. Thus Paul Harris has written of a "cancer of Westphalia," referring to the treaty of 1648 that ended the Thirty Years War and is generally, if somewhat mistakenly, held to have initiated the modern sovereign state. In his view and that of others, this must be overcome by international governance, and by "people-centred diplomacy," which will involve including democratically chosen representatives of people most affected by climate change in international negotiations.[5]

Environmentalists who write in this way could not possibly be more mistaken. They offer no practical advice whatsoever about how state elites can be persuaded to surrender power in this way, or how popular representatives can be "democratically chosen" in countries with weak or non-existent democracies, or how and why sovereign peoples would accept being outvoted or dictated to by others. Even if it could by some miracle be introduced, such international governance would lack the popular legitimacy to take any effective actions. The nationalist backlash against the European Union should be a sufficient lesson in that regard. Such empty recommendations are a major distraction of ecological and progressive thought and effort away from practical politics and the achievement of practical goals.

As Dani Rodrik has written, "Global governance has a nice ring to it, but don't go looking for it any time soon. Our complex and variegated world allows only a thin veneer of global governance—and for very good reasons too."[6] So nation states are not going away; and if climate change and other challenges are to be met, then the states of the 21st century will have to be strong and to achieve a great deal; and this is especially true of those developed and developing states that are responsible by far for the greater part of CO_2 emissions.

The greatest source of a state's strength is not its economy or the size of its armed forces, but legitimacy in the eyes of its population: that is, a general recognition of the state's moral and legal right to authority, to have its laws and rules obeyed, and to be able to call on its people for sacrifices in the form of taxes and when necessary conscription.[7]

Without legitimacy, a state is doomed either to weakness and eventual failure, or to becoming what has been called a "fierce" state, ruling by fear. Such states have the appearance of strength, but they are inherently brittle and liable to collapse if people cease even for a day to be afraid of them—as several Middle Eastern rulers discovered in 2011. The basic weakness of the European Union compared to its member nations is that it has never achieved real legitimacy as a quasi-state authority in the eyes of most Europeans.

There are many different sources of legitimacy. One of them is simply to have been around for so long that a given state appears to be part of the natural order of things—what Max Weber called traditional legitimacy. This, however, can be lost if society and the economy change in ways that render the state archaic (as with the ancien régime in France in the decades before the Revolution). Another obvious source of legitimacy is performance, or success at whatever tasks the population thinks are truly important. Some of these state goals have remained constant over time: defense against external enemies and the maintenance of basic internal security have been tasks of the state since states first came into existence.

Others have changed over time. In the West at least, the correct observance of religious laws is no longer significant outside certain parts of the United States. On the other hand, all states are now expected to provide social welfare, health services, and education to their people—a goal that hardly existed 200 years ago; and their legitimacy depends in part on their success at this. As already noted, the management of water—the prevention of floods and droughts—has always been a key state goal and source of state legitimacy in China, but until recently it was a minor issue in most of the West. If the struggle to limit climate change is to be won, then in the decades to come, populations will have to be convinced that success in limiting CO_2 emissions is a vital state goal and source of state legitimacy.

This will take time, and at the same time, for reasons set out in Chapter 2, populations will be coming under increased social, economic, ecological, cultural, and demographic pressure. One of the ways of understanding legitimacy is that it buys time. In other words, a system that enjoys other sources of legitimacy can afford to fail for longer at various tasks than a state lacking such legitimacy. One important source of legitimacy over the past 70 years has been democracy, leading to the toleration of failures by elected governments and the acceptance by minorities of majority votes (or, remarkably, in the United States, the acceptance by majorities of minority electoral victories, because this is allowed by the Constitution).

But as a whole range of democratic and semi-democratic states have discovered over the past century, democracy alone will not preserve a given state indefinitely if that state is deeply divided internally and fails to achieve what the population sees as vital goals. For this, a deeper source of legitimacy is necessary, rooted in a common sense of national belonging. In the modern world, the greatest and most enduring source of this feeling and this state legitimacy has been nationalism.

Except for communism during its brief revolutionary periods, nothing in modern history has compared to nationalism as a source of collective effort, sacrifice, and, of course, state construction. Other elements of an individual's identity may be important in personal terms, but (with the exception of the Muslim world, where religion remains strong) they do not build great and enduring institutions. The small minority of permanent expatriates who juggle different identities as they move from country to country forget that in every country they live in, they are dependent on the state for protection and support—and strong states were generally put together in the past with the help of strong nationalisms.[8]

Suggestions that something like loyalty to a football team is any sort of equivalent of nationalism are foolish. With the exception of a few hooligans, nobody has ever given his or her life for a football team. Nor is it really of much consequence whether a football team continues to survive after its existing generation of fans are dead.

In Russia, it was above all the revival of Russian nationalism that pulled the country back from complete collapse in the 1990s. In China, as state economic strategy has changed into something like authoritarian social market capitalism, nationalism has replaced communism as the legitimizing ideology of the state. This may also prove to be the case in the West, as liberal democracy fails to achieve its goals of social equality and growing prosperity for the population as a whole. Just as the communist state continues in China but with a nationalist content, so democracy may survive in the West but with nationalism replacing liberalism. As of 2019, this process seems to be well underway in some countries of the EU.

Nationalism and Conflict

At the core of fears about nationalism over the past hundred years has been the belief that nationalism produces war. The prevention of international war has been the chief purpose of all liberal internationalist projects since Immanuel Kant wrote his *Perpetual Peace: A Philosophical Sketch* in 1795. The association between nationalism and war seemed firmly established by the ghastly experience of the two world wars, as well as numerous lesser conflicts in Europe. The League of Nations, the United Nations, and the European Union, as well as other less successful regional groupings, were all created to prevent further international wars.

Since 1945, however, the world pattern seems to have changed quite significantly.[9] The number and scale of international wars has greatly diminished. Direct war between the great powers has been strongly discouraged by the development of nuclear weapons; and if the United States and the USSR could avoid nuclear catastrophe, there seems reasonable hope that the great powers of the 21st century will be able to avoid it also. The possession of nuclear weapons by Pakistan since 1998 seems the obvious reason that India has not reacted to terrorist attacks by Pakistani-based groups by going to war with Pakistan. Given India's superiority in conventional forces, in a previous era war would almost certainly have resulted.

What we have seen since 1945 is a very large number of civil wars and rebellions, sometimes with outside great powers becoming involved on one side or another. More than 90 percent of wars over the past 70 years have been internal. Occasionally these have been motivated in part by state nationalism. In the cases of Korea and Vietnam, these were wars for the reunification of countries divided by the Cold War. In a majority of cases, however, the civil wars have been the result of state weakness and collapse, stemming in large part from the failure to generate a strong sense of state nationalism (though these wars are often caused by separatist ethnic nationalism). This casts the question of the relationship between nationalism and war in a rather different light from the way in which it is usually portrayed in liberal internationalist circles.

In Western countries since the end of the Cold War, it has been the liberal internationalists, in league with American imperialism, who have done most to stoke international conflict.[10] In Europe, it was the supreme liberal internationalist, Tony Blair, who led Britain into the Iraq War. Hillary Clinton backed the Iraq War and initiated the disastrous overthrow of the Libyan state. In the United States, there is no significant difference between the Republicans and the traditional Democratic establishment when it comes to confronting China and threatening the overthrow of the Chinese state by "democratic" subversion, while the Democrats have been just as responsible as the

Republicans for driving hostility to Russia, and after 2016 tried to whip up anti-Russian fury as a weapon against Trump.[11]

The US foreign and security establishment (or "Blob" as Obama's adviser Ben Rhodes dubbed it) that is willing to engage in repeated wars to maintain US global hegemony is thoroughly bipartisan. As a result, it gets cover from both Republican and Democratic media and from Congress; and so its members have remained in place and unpunished despite repeated and sometimes disastrous mistakes and failures over the past generation.[12]

In Western Europe, nationalism in the early 21st century, while often ugly, is also profoundly *defensive*. As Yael Tamir has pointed out, contrary to the rhetoric of some European leaders, even the breakup of the European Union (deeply undesirable though this would be) would not lead to new wars between France, Germany, and Poland. Today, and for a long time to come, such wars between them are simply unimaginable given their demographics and their cultures (especially their youth cultures).

This applies just as much to the national right as to liberals and the left. The Front Nationale is not dreaming of conquering the Rhineland for France. Alternative für Deutschland is not planning a war against Poland to recover Silesia. The Swedish Democrats are not interested in making Sweden a military great power. On the contrary, these parties have often opposed international actions and commitments because of their desire to concentrate on what they take to be a far greater threat—the danger posed by mass migration to the social and cultural integrity of their societies.

Nationalism and Reform

Apart from the issue of nationalism and war, belief on the part of environmentalists that the erosion of nationalism and the nation state is a good thing is based on a triple misconception concerning the relationship of nationalism to modernity, to globalization, and to social, economic, and cultural progress.

This book is not the place to delve into the vexed question of the ancient roots of modern nationalisms. What can be acknowledged by all sides in this debate is that in the two centuries following the French

Revolution, nation states and nationalisms became the frames of modernity, first in Europe and the United States, then in much of the rest of the world: "Modern political life has an inescapably national dimension to it."[13] This link between nationalism and modernity is in fact the thesis of the "constructivist" theory of nationalism, even if its proponents differ on the dates and the precise socioeconomic configurations that produced modern nationalisms.

But as Tom Nairn and others have pointed out, the members of this school have shied away from considering the logical corollary of this view, namely, that nationalism and the nation state have been, and remain, the "constitutive principle of modernity";[14] that if religiously sanctioned monarchical rule over fractured territories was no longer a viable state form politically or economically, then the only alternative is a polity based on the sovereignty of a bounded national citizenry over a bounded territory, held together by common national sentiment. This is also the only form in which effective democracy can be organized.[15]

The link between nationalism and modernity is equally clear in reforms instituted by Asian countries that in the 19th and 20th centuries sought to modernize in order to defend themselves against the Western imperial powers and gain for themselves an active rather than passive role in capitalist globalization. The adoption of capitalism and Western social and cultural models was an immensely painful and contested process. Ancient cultural traditions had to be abandoned, social and political hierarchies overturned, and everything from dress to the regulation of sexual relations radically transformed.

Quite apart from the cultural impact, these changes imposed great physical hardships on ordinary people, who were expected to pay higher taxes to build modern infrastructure, to accept conscription into new mass armies, and to submit to being driven from their farms into urban slums to make way for a new international commercial agriculture. The moral, political, social, and economic sacrifices involved were colossal—as they are likely to be in the struggle against climate change and in adaptation to artificial intelligence. Not surprisingly, therefore, these reforms met bitter resistance; and also not surprisingly, across most of Asia, they failed. As Tom Nairn writes of the European 19th century, in words that are equally applicable to the great waves of capitalist social change sweeping Asia today: "Urbanisation is the smooth-sounding, impersonal term for what was often an agonising process: the fearful undertow of modernity. During it, rural emigrants look backwards as much as forwards, and pass from the remembrance to the often elaborate reinvention of the worlds that they have lost."[16]

It can be confidently stated that the only countries where these reforms succeeded were those where the state was able to mobilize a strong sense of united nationalism as a justification for the sacrifices involved. They were presented, with conviction, as necessary to strengthen the nation against the threat of alien conquest or domination. Japan is the preeminent example of this successful strategy in Asia, as is Kemal Ataturk's Turkey in the Muslim world.

The astonishingly radical reforms carried out under the Meiji regime in Japan from the 1860s on were explicitly justified and legitimized by the need to strengthen the nation and avoid the fate of other Asian countries at the hands of European imperialism. As in Turkey, the reforms were carried out by men with a military background or with a strongly military cast of mind. The official slogan was not, as in 19th-century France, "enrich *yourselves*" (*enrichissez-vous*) but "enrich the country, strengthen the army": "Each modernization effort was clearly related to the central problem of increasing the wealth and power of the [Japanese] nation, and almost every major move was initiated and pushed by the national state in order to serve clearly defined national aims."[17]

At the core of these reforms and their legitimization was the spread of a new kind of Japanese nationalism through a new mass education system.[18] This nationalism was not "created" but reshaped and extended on very old foundations. Indeed, a preexisting nationalism and the emperor as a universally accepted (if previously entirely symbolic) source of state legitimacy were crucial to the success of the entire reform process.[19] In pre-Meiji Japan, under the Tokugawa Shogunate, "The emperor was popularly accepted as the ultimate source of all political authority. . . . Moreover, there was a widespread sense of national commitment. The identification with Japan as a culture and a nation had repeatedly surfaced with a strong unifying potential whenever the nation was externally threatened."[20]

As a Japanese *liberal* reformer of the 1880s wrote, "The one object of my life is to extend Japan's national power. Compared with considerations of the country's strength, the matter of internal government and into whose hands it falls is of no importance at all."[21]

Liberal capitalist reform in the developing societies of 19th-century Europe also depended critically on nationalism for its legitimation.[22] Indeed, 19th-century liberalism was intrinsically connected to nationalism, and European liberalism's break with nationalism after the First World War marked a radical departure from its origins. And while liberal free-market reform was meant to reduce the power of the state over the economy, it depended on state power to beat down all those forces opposed to reform. As Immanuel Wallerstein has noted, "Liberalism has always been in the end the ideology of the strong state in the sheep's clothing of individualism; or to be more precise, the ideology of the strong state as the only sure ultimate guarantor of liberalism."[23]

Liberalism as it grew up in early and mid-19th-century Europe was closely associated with movements of national "liberation" and/or reform for the sake of national strength in the face of imperial domination by other states. In Britain too, the father of modern liberalism, John Stuart Mill, closely associated liberal progress with the creation of strong and homogenous national states.

Nationalism was at the heart of what Antonio Gramsci later called the "hegemony" of bourgeois liberal ideas—including capitalist economic reforms—in late 19th-century Italy: their acceptance by much of the population as a form of "common sense." This helped bring about the consent of a majority of the population, most of the time, to elite rule and elite policies of reform, even when these clearly did not serve the short-term interests of the people concerned.[24] In many parts of Europe, to ram through deeply painful liberal reforms required that 19th-century liberals be openly elitist (including advocating very restrictive franchises), frequently authoritarian, and appealing to nationalists, as nationalism was the one force that could bind enough of the population to the liberals to support liberal reforms.[25]

Contemporary liberal reformers in the EU and elsewhere have retained the elitism and even the authoritarianism of their 19th-century predecessors, but they have forgotten the nationalism. This elitist authoritarian strain was much in evidence in Russia in the 1990s. While parading to Western audiences their commitment to "democracy," the liberal intelligentsia of Moscow and St. Petersburg were entirely open in their contempt for ordinary Russians. They called them in almost racial terms "Homo Sovieticus," and their attitudes to them did indeed resemble the racist attitudes of the northern Italian elites to the conservative southern Italian peasantry after Italian unification, or the white Latin American elites to the darker-skinned masses in their countries.[26]

In recent years, such anti-democratic attitudes have resurfaced among liberals in Europe and North America—very much along 19th-century liberal lines—in reaction against illiberal tendencies like the Brexit vote and the mass movement in support of Donald Trump. As in Russia in the 1990s, liberals were even reckless enough to display this contempt publicly while asking the masses for their votes (like Hillary Clinton with her talk of "deplorables").[27]

The mistake of the liberal reformists of the present era has been not to understand that the only way their 19th-century predecessors were able to persuade the masses to accept their rule and their program was through the appeal of nationalism. In American history, as the historian Michael Kazin has written, it is nearly impossible to name "any American radical or reformer who repudiated the national belief system and still had a major impact on US politics and policy."[28] Jill Lepore adds that "In American history, liberals have failed, again and again, to defeat illiberalism except by making appeals to national aims and ends."[29] This failure has been especially disastrous in the case of the Russian liberals, who willingly allowed themselves to be cast not only as exponents of a horribly painful program of economic reform but also as defenders of US hegemony over Russia—not a good election platform as far as most Russian voters are concerned.

I got very tired during my years in Russia in the 1990s of hearing from Western analysts and some Russian liberals that the 19th-century Russian "Westernizers" were precursors and models for contemporary Russian pro-Western reformists who believed in Russia becoming a subservient ally of the United States. The 19th-century Westernizers certainly believed in liberal reforms for their own sake. Like their equivalents in China or Japan, however, they also believed that these reforms were necessary to strengthen the Russian Empire in its competition with its West European rivals. The idea of carrying out reforms so that Russia could become a client state of the British Empire would not have occurred to their minds.

The behavior of Arab liberals today in supporting authoritarian rule, out of both their fear of the conservative Muslim masses and their hope that authoritarianism provides an avenue to reform, is therefore completely in keeping with 19th-century liberalism. What has crippled them is the perception that they are acting as US vassals. If there was ever any chance of the military regime in Egypt conducting a successful program similar to the "Kemalist" authoritarian reform carried out by Ataturk in Turkey in the 1920s and '30s, they lost it when Anwar Sadat made peace with Israel and agreed to make Egypt a client state of the United States.

This is the problem that reformers all over the Muslim world have faced when trying to adopt Kemalism as a model for their own societies. Ataturk's radical Westernizing reforms were legitimized not only by nationalism but also by *victorious* military nationalism.[30] He made his own name as an Ottoman general by beating the British imperial forces at Gallipoli. In 1919–22 his Turkish nationalist forces beat not only the Greeks and the Armenians but the Western Allies as well.[31] Having defeated the West, Ataturk had the nationalist legitimacy to imitate the West.

There is an interesting contrast here with the Pahlavi dynasty in Iran, which tried to implement many of the same reforms as Ataturk (Reza Shah even mimicked Ataturk in banning traditional dress), but as clients first of the British and then of the Americans, they lacked the nationalist legitimacy to gain mass acceptance of their reforms. In Pakistan, three leaders (two military and one civilian) have explicitly declared Ataturk as their model and inspiration for the reform and development of Pakistan: General Ayub Khan, Zulfikat Ali Bhutto, and General Pervez Musharraf. They all failed, above all because given underlying Pakistani realities, they were unable to generate anything like a united nationalist movement in support of their programs.

The greatest apparent triumph of liberal internationalist reforms after the end of the Cold War also owed far more to nationalism than liberals have been prepared to recognize. The accession of the former communist states of Eastern Europe to the European Union meant obeying orders from Brussels to accept reforms that many East Europeans found absolutely detestable. In Lithuania, where I was stationed as a journalist for the *Times* (London) in 1990–92, almost everything that the EU was demanding was alien to Lithuanian conservative nationalists. Partly because of enforced Soviet isolation, partly because of their own traditions, their natural instinct was to return to the values of Lithuanian nationalism during the period of Lithuania's independence—in other words, before 1940. As Anne Applebaum writes of a similar Polish nationalist supporter of the Law and Justice Party in Poland, "She never turned against liberal democracy, because she never believed in it, or at least she never thought it was all that important. For her, the antidote to Communism is not democracy but an anti-Dreyfusard vision of national sovereignty."[32]

It was hardly surprising, therefore, that Vytautas Landsbergis, the Lithuanian leader, proposed as the motto of the new Lithuania "Work, Family, Fatherland"—the motto of Vichy France, though he doubtless did not realize this. For Lithuanians of my acquaintance in 1990–93, the idea of gay marriage, for example, was so monstrously weird that it evoked not horror but sheer blank incredulity.[33]

The economic changes involved in the East European transition from Communism to the free market were equally wrenching, involving the destruction of much of industry and the impoverishment of much of the countryside through the entry of subsidized EU agricultural products. The badly paid but secure jobs and much of the basic social welfare guaranteed under communist rule both vanished. Eventually, these changes paid off as far as most of the populations were concerned, but many areas and groups (especially the elderly) have never really recovered.[34]

In the late 1990s I worked as a correspondent for the *Financial Times* based in Budapest; I witnessed how central parts of that and other cities were blooming while the industrial suburbs and much of the countryside sank into poverty and social despair. As with all past capitalist revolutions and with modern Western capitalism in general, the benefits of capitalist growth were very unevenly distributed, but with a malign post-communist twist: so many of the elites who benefited from the change to capitalism came from the former communist elites, and as in Russia, their new ascendancy under capitalism came from privatizing former state property into their own pockets.

How then were these populations persuaded to accept these changes? Part of the answer is, of course, the Western ideological, cultural, and economic hegemony that replaced communism. The other part, however, comes from nationalism. They accepted this process because they were told that if they did not, they would never get into the EU and NATO (membership in which essentially required membership in the EU as well); and sooner or later they would fall back under the hated domination of Moscow.[35]

There is nothing surprising about the current ascendancy of populist conservative nationalism (combined with increased welfare policies) across much of Eastern Europe. This represents not "backsliding" but continuity.[36] Now protected by NATO, these populations have no real reason to fear Russian conquest. Instead, the EU itself is widely seen as a quasi-imperial force, attempting to dictate (through the acceptance of Muslim refugees) demographic changes that strike at the core of East European ethnic nationalist beliefs. As a Hungarian friend (who in terms of economic reform and most social attitudes considers himself a liberal) remarked to me, except for the expulsion of Germans in 1945–46, even the Soviet hegemons never dared to change the ethnic composition of East European societies, whatever they may have done within the USSR itself.

Nationalism and the Transcendence of Time

Apart from nationalism's legitimization of painful reforms in general, what especially brings nationalism and ecological thinking together is the capacity of both to demand sacrifices on the part of existing people for the sake of future generations.[37]

The fact that this link between nationalism and the struggle against climate change is almost never mentioned by writers who have examined the issue of climate change and "intergenerational justice" is in itself testimony to the monolithic hostility of liberal establishments to nationalism.[38] This is particularly odd because when talking about intergenerational responsibility and solidarity, many contemporary left-wing environmentalists are beginning to sound very much like Edmund Burke. In the words of a writer in the left-wing American journal *The Nation*:

> Every person, whether or not they have children, exists as both a successor and an ancestor. We are all born into a world we did not make, subject to customs and conditions established by prior generations, and then we leave a legacy for others to inherit. Nothing illustrates this duality more profoundly than the problem of climate change, which calls into question the very future of a habitable planet. . . . To combat the apocalyptic apparitions, we need to conjure alternative worlds, leaping forward *and* looking back.[39]

A great many national rituals are specifically designed to evoke this sense of organic continuity through time, and their performance by successive generations does eventually give them a certain timeless quality.[40] This sets both nationalist and ecological thinking against dominant trends in contemporary culture and economics. Christianity also has the idea of sacrifice for others at its very core; and ecological thinking is playing an increasingly large role in the thinking of a number of churches, most notably the Roman Catholic Church.[41]

It was only when I began to read approaches to climate change by mainstream economists that I came fully to understand the extent of our moral decadence as a culture. A great many of these approaches are written from the standpoint that the interests of future generations matter little, or not at all.[42] Thus 6 percent is a common "discount rate" among economists when it comes to valuing benefits for future generations against those of the present generation.

As Nicholas Stern has pointed out, this would mean a "unit of benefit" in 50 years being valued 18 times lower than it is now, and in 100 years 339 times lower: "To assume such a rate comes close to saying 'forget about issues concerning 100 years or more from now.'"[43] Or to put it in more individual terms, "I'm all right, Jack—my children and grandchildren can go to hell." And remember: while children and teenagers alive today will not live to experience the most catastrophic effects of climate change, we can already see, from the increase in heat waves and wildfires and the destruction of nature, that the world they will live in before they are old will be a severely diminished one compared to the one we have enjoyed. Most (though by no means all) economists are mired in the contemporary, and they seem only able to focus on the issue of how to get people to keep spending *now*.

This is their usefulness and why they are needed by business and by government—but it makes them simply inappropriate as modelers of a future society with quite different requirements.

Indifference to the future is deeply rooted in contemporary materialist and individualist culture, but it is a morally criminal position for any parent to take, at least parents who like to think of themselves as decent and responsible.[44] Indeed, it suggests that there is no point in having children at all—which indicates a relationship between such thinking and the declining birth rate in developed countries. Such an attitude is also antithetical not just to nationalism but to the very idea of a nation.

The destruction of famous national landscapes and the loss of the historic capitals of England and the United States to sea-level rise after present generations are dead is of little concern to a materialist, but of immense concern to a nationalist who associates his or her identity with national traditions. Without such concern to pass on national inheritance, as Edmund Burke wrote, "No one generation could link with another. Men would become little better than the flies of summer."[45]

A considerable problem for the mobilization of action against climate change is that making people look at a future of climate change is liable to induce despair and a retreat into narcissistic self-indulgence. The motto of climate change activism, therefore, should be Gramsci's maxim: "Pessimism of the intellect, optimism of the will." It is very difficult to generate optimism of the will, however, without a measure of hope. And in view of our mortality, there is no long-term hope for individuals—we're all dead. As far as most people are concerned, existential hope can only be invested in descendants and in ongoing communities.[46] The following passage from Tom Nairn has special meaning in the context of environmental destruction and the multigenerational struggle against future climate change:

> Walter Benjamin wrote that the appalled Angel of History, who seems to be contemplating in dismay modernity's piles of wreckage upon wreckage, "would like to stay, awaken the dead, and make whole what has been smashed." But through nationalism the dead are awakened, this is the point—seriously awakened for the first

> time. . . . Through its agency, the past ceases to be "immemorial": it gets memorialised into time present, and so acquires a future.[47]

The tremendous emotional power of the nation stems largely from the way in which it can echo loyalty to the family and appeal in cultural terms to that most fundamental of all instincts, the desire of the living organism to propagate its genes.[48] This has always set conservative nationalists at something of a tangent to radical free-market liberals, and today it brings them closer to the moral and philosophical underpinnings of environmental thinking. Nationalism is rooted in a sense of national society in Burkean terms as a covenant between the dead, the living, and those yet to be born—a sentiment close to the environmentalist maxim that "the world is not given to us by our fathers but borrowed by us from our children." This in turn is related in complicated ways to the human sense of the sacred.

The idea of a nation thinking of itself as living for only one generation is a contradiction in terms; and this is true not only of those nations founded (whether accurately or not) on the idea of ancient ethnic identity, or those founded on some version of ideology. The central monument to the idea of past wars in almost all countries is called the *eternal* flame. The motto of the Great Seal of the United States does not read, "The US Constitution is so, like, ten minutes ago." Drawing on Virgil's vision of the destiny of Rome, it reads *Novus Ordo Seclorum*: "A New Order of the Ages."[49] The idea of the permanent existence of sovereign nation states is also fundamental to international law and the modern international order.[50]

It is undeniable that there is a very deep gulf between this kind of thinking and the greater part of contemporary culture in the countries of the developed world. Populations live in the "eternal present" of the social media and the news cycle. Economies depend on the relentless promotion of short-term gratification, the creation of "artificial needs" through advertising, and the resulting constant replacement of possessions with new ones generated by fashion—all of which has to be paid for in terms of higher carbon gas emissions and other environmental damage.

This is becoming as true of China and India as it has been in the West, Japan, and South Korea for several decades. Chandran Nair describes the modern informal festival of "Singles Day" in China (November 11, or 11/11 in the Western calendar), when Chinese youth celebrate their unmarried condition and (for a day at least) lack of responsibility to their families, as having become "the largest shopping day in the entire world," thanks to relentless promotion by Alibaba and other Chinese e-commerce corporations. In 2017, Alibaba alone sold $25 billion worth of goods that day, contributing to making it "probably the biggest single organized mass-shopping contributor to carbon dioxide emissions the world has ever known." Changing this kind of culture across the planet is going to take a tremendous effort.[51]

Faced with this profound cultural obstacle, a good deal of thought is now being devoted to how to change modern culture and shift it toward a concern for the long term. This has led to efforts like the Future of Humanity Institute at Oxford, the Long Now Foundation, and the Long Time Project. European parliaments now have committees on future thinking and future challenges. There has been a rather desperate and strained search for historical images and models for such cultural approaches. These include the European cathedrals of the Middle Ages, most of which were famously built over several generations and even centuries at a cost to local societies almost inconceivable for the present era, and the alleged "seven generations" thinking of the Iroquois people of North America.[52]

These images, however, reveal the full extent of the challenge facing those who want to change contemporary culture in this way. The medieval cathedrals were the fruit both of a deeply and universally held religious faith and culture (enforced when necessary by ferocious sanctions) and of local urban communities that had taken shape over hundreds of years and were characterized by an intense sense of local identity and loyalty. The difference with the great multi-cultural urban agglomerations of the contemporary West could hardly be more stark.

As to the Iroquois (who did not formally have anything called "seven generations" thinking, though it is possible that they thought along these lines), they are famous both for the sophistication of their internal tribal political practices and the extreme ferocity of their attitudes and behavior toward other peoples, legitimized by their religion.[53] In other words, this kind of long-term thinking was the product of old cultural identities based on sacred beliefs held by populations as a whole.

As Prasenjit Duara has written, in words that apply to a large part of ecological thinking:

> Modern universalisms have tended to lack confidence in investing the transcendent or utopian truths they propose with symbols and rituals of sacred authority. Their hesitation doubtless has good reason that we may see from the rampaging power of extreme nationalisms, such as Nazism, or blinding faith in utopian science triumphant over reason. *But no movement of major social change has succeeded without a compelling symbology and affective power.* [my italics][54]

Reflecting this challenge but also their own sense of the sacredness of nature, a good deal of ecological thinking is characterized by a desire for deep cultural transformation linked to religious impulses (one aspect of which is "Deep Ecology" thinking) and the need for a new "Axial Age" in which the moral basis of human civilization will be transformed.[55]

Such religious ecological ideas may well spread widely in the future, driven by collective shame and nostalgia for a natural world that we have destroyed, and related to a culture of austerity enforced by reality but legitimized by religion. If a heretical sect of a rebellious religion founded by a carpenter from the fringes of the Roman Empire could end up by becoming the religion of that empire and perpetuating its cultural legacy, then anything is possible.

But the span from the death of Christ to the conversion of Constantine was 282 years; and as every reputable climate expert is shouting at us, we have only a fraction of that time to take much more serious action to limit climate change. Even if our present civilization does resemble the later Roman Empire in its decadence, it is still our duty as citizens not to allow it to be destroyed in the vague hope that some morally better culture may one day come along to replace it.

In the world of today and probably for a very long time to come, the only truly popular cultural force that retains this mass appeal and capacity for long-term thinking is nationalism. Islam might have done the same in the Muslim world, but it is still engaged in working through its relationship to modernity. With the possible exception of Iran, where religion combines with a powerful and ancient national identity, Islam seems likely to be obsessed by this internal struggle for a long time to come and thereby neutralized as a force capable of shaping modern culture in a positive direction.

Nationalism's ability to project its thinking into the future is closely related to its ability to draw on the past (whether real or re-imagined)—what Anthony Smith called the "national myth-symbol complex."[56] This is in turn largely responsible for nationalism's ability to inspire effort and sacrifice. Garrett Mattingly wrote of the English legend of the defeat of the Spanish Armada, in the aftermath of the British fight against Nazi Germany in the Second World War: "It raised men's hearts in dark hours, and led them to say to one another: 'What we have done once, we can do again.' In so far as it did this the legend of the defeat of the Spanish Armada became as important as the actual event—perhaps more important."[57]

Since then, it has been above all the image of the Second World War—or mostly the year from June 1940 to June 1941 when Britain fought almost alone—that has obsessed the British public memory.[58] As a historian, one must deplore the mythological elements involved; and this legend played its part in creating the disastrous illusions that underlay the vote to leave the European Union in 2016.

On the other hand, as Chapter 4 will point out, the legend of British national unity and social solidarity during the war is also a progressive image, closely linked to the creation of the modern British welfare state. It is of importance to the struggle against climate change because it is an image not only of collective military sacrifice but of willingly accepted austerity and rationing for the sake of the nation and the common good (and if it was not always really so willing—well, this is a legend, remember).

Nationalism's attachment to the past (once again, whether real or re-imagined) has a direct connection to environmentalism through a concern for the preservation of the national landscape and natural heritage. Thus in a survey of 1997 by researchers at Nuffield College Oxford, 84 percent of Scots and 72 percent of Welsh put the countryside first among focal points of their nationality.[59] As Theodore Roosevelt wrote,

> Defenders of the short-sighted men who in their greed and selfishness will, if permitted, rob our country of half its charm by their reckless extermination of all useful and beautiful wild things sometimes seek

> to champion them by saying the "the game belongs to the people." So it does; and not merely to the people now alive, but to the unborn people. The "greatest good for the greatest number" applies to the number within the womb of time, compared to which those now alive who form but an insignificant fraction. Our duty to the whole, including the unborn generations, bids us restrain an unprincipled present-day minority from wasting the heritage of these unborn generations.[60]

A mystical attachment to nature and place has ancient roots in human culture: "The Apache believe that wisdom sits in places, and indigenous groups everywhere, from Amazonia to British Columbia and the mountains of Taiwan, celebrate their long-standing and unbreakable bond with the land wherein they dwell."[61] In the West, a passionate cultural attachment to nature reappeared in the 19th century both as a result of Romanticism and in reaction against the radical transformation and widespread destruction of nature by the Industrial Revolution.[62] A love of place and nature is far older. It permeates the works of Virgil and Horace as—a few thousand miles and centuries away—it does those of the T'ang poets Li Bai, Du Fu, and Po Chu-I. Not least among the effects of climate change on human societies may be the destabilizing psychological effects of radical changes to weather patterns, landscapes, plants, and animals felt to be part of the very nature of life.[63]

An appeal to this aspect of nationalism is viewed with special suspicion in liberal and left-wing environmentalist circles because an extreme form of it was part of the cultural basis for "blood and soil" nationalism, and ultimately for Nazism and the Nazi pseudo-religion of neo-paganism. It is appealed to in these terms by a limited number of extreme right-wing figures today.[64] To use this, however, as a reason to reject the entire tradition of attachment to local and national landscapes in the West is to throw out the bathwater, the baby, the bath, and the entire municipal drainage system.

It is another example of a failure to prioritize climate change as opposed to narrow ideological agendas; and since the vast majority of people on the right who are deeply attached to local nature are not fascists but traditional centrist conservatives, it also involves yet another refusal to seek support for action against climate change across party lines.[65] It is also hardly logical or morally viable to celebrate the attachment of indigenous people in India or Australia to the defense of sacred natural landscapes while denying the same right to German or French people.

In this as in mobilization against climate change in general, ecologists would do better to listen to Edmund Burke: "To be attached to the subdivision, to love the little platoon we belong to in society, is the first principle (the germ as it were) of public affections. It is the first link in the series by which we proceed towards a love to our country and to mankind."[66] In other words, concern for the future of the national environment provides an essential link between purely local movements of conservation and the struggle against climate change by nation states in cooperation with each other—without, as I have argued, *replacing* action by those states.

What Kind of Nationalism?

The question is not whether nationalism is important to states, but what kind of nationalism can best strengthen Western democratic states so as to meet the challenges of the 21st century, climate change first among them. If, as Chapter 2 argued, the idea of a borderless state with a completely open identity is a path to disaster, then an attempt at re-creating a state based on a narrow and closed ethnic and cultural identity is obviously not desirable for any country containing large ethnic or religious minorities. It is also not achievable, without at the very least ruthless repression, at the worst ethnic cleaning, deportation, or genocide. So the answer would appear to be civic nationalism with a strong cultural and ideological core, based on a much stronger idea of common citizenship giving a common sense of identity to all citizens, of all races, genders, and classes:

> [Liberals] must offer a vision of our common destiny based on one thing that all Americans, of every background, actually share. And that is citizenship. We must relearn how to speak to citizens *as citizens* [italics in the original] and to frame our appeals—including

> ones to benefit particular groups—in terms of principles that everyone can affirm. Ours must become a civic liberalism.[67]

In the contemporary West, conservatives need to recognize that the kind of nostalgic dream of an integral, racially unitary nation represented by much of the vote for Brexit and Donald Trump is just that—a dream. Any attempt at the re-creation of such a society will in the end point toward fascism. Equally, however, Western liberals need to recognize that the bonds of national citizenship are essential to many basic liberal causes, that no country if it wishes to survive can be completely open in terms both of borders and cultural identity—and to admit that liberals themselves in fact set high cultural standards for real integration. If Western democratic societies are to build the national unity and resilience that they will need to survive in the years ahead, then conservatives need to become a great deal less ethnic nationalist, and liberals need to become a great deal more civic nationalist—but the political spectrum as a whole needs to be nationalist.[68]

The United States has been famously open to immigrants (European ones, that is, until the 1960s), with a semi-official identity based on ideological and constitutional, not ethnic, loyalty (to the so-called American Creed). This has certainly not made American nationalism a weak and diluted brew. It has, however, required an extremely high degree of ideological and cultural shaping through the educational system, the media, and popular culture (especially Hollywood) aimed both at assimilating the immigrants and gaining acceptance of them by the existing population. It didn't just happen.[69]

The term "American nationalism" came into being toward the end of the 19th century, and it was inculcated through the schools and public symbolism, very much as a way of binding together in a strong common loyalty and identity the older (white) population and the new (white) immigrants. As Chapter 4 will argue, far from being a reactionary force, this nationalism was closely associated with the Progressive movement aimed at taming the wild capitalism of the "Gilded Age," creating basic social welfare and modernizing the federal government.[70] It was the basis for Theodore Roosevelt's platform when he sought re-election to the presidency as an independent, and it later provided the foundations for Franklin Delano Roosevelt's New Deal.[71]

It would therefore be a mistake to see civic nationalism as necessarily multi-cultural. For one thing, in the past, civic nationalism was just as insistent as ethnic nationalism on adoption of the national language as a criterion of citizenship. John Stuart Mill is considered the greatest prophet of liberal individualism, but this did not extend to the toleration of different national identities within the state: "Free institutions are next to impossible in a country made up of different nationalities. Among a people without fellow-feeling, especially if they read and speak different languages, the united public opinion, necessary to the working of representative government, cannot exist."[72]

The civic nationalisms of America and France in particular have had a strong element of what I have called "civilizational nationalism," a belief that their systems represent the best, or even the only true, version of civilization for all of mankind. The mainstream left in France, while friendly to immigration in principle, has been especially insistent on the need for everyone to accept "Republican values"—ones that are in fact imbued with a particular culture.

Similarly, most Western cultural liberals, while genuinely open in terms of color of skin, and theoretically open in terms of culture, are in fact dogmatic absolutists and not cultural pluralists when it comes to their own core moral and cultural attitudes (which, just like the ancient Romans and Chinese and the religious fundamentalists, they see as the only universally valid moral norms); and while declaring publicly that "no voice has a privileged status," they are in fact implacably convinced that one voice and only one should have such status: their own.

This is especially true in the areas of women's rights, feminism (including something like a cultural obligation to seek paid employment), LGBTQIA rights, marriage, abortion, female genital mutilation, and sexual relations in general (mostly quite rightly, in my view).[73] Liberal approaches to these issues as far as conservative immigrant communities are concerned could be seen as the liberal form of compulsory assimilation; and these are also the areas where moral and cultural imperatives tend to be at their strongest and their most deeply shaping as far as societies are concerned.[74]

This liberal approach is a thoroughly incoherent one for it crashes against the other liberal principles of multi-culturalism, the acceptance of separate ethnic and ethno-religious communities within the state, and open immigration. In fact, the European liberal approach to immigration might be unsympathetically summed up as allowing as many conservative Muslims as possible into Europe and then infuriating them as much as possible once they are there.

As Walter Russell Mead, Michael Lind, myself, and others have noted, for most middle-class Americans, acceptance into the American community has never just required acceptance of American political values and respect for the Constitution and the rule of law.[75] It has also involved a whole set of explicit or implicit cultural requirements, including knowledge of English, American patriotism, the "work ethic," family, religious faith (even if the definition of what that faith should be has changed over time), and some understanding of extremely nationally specific, arcane, and complicated American popular rituals like baseball. And it worked. White middle-class America successfully inculcated this culture into tens of millions of European immigrants and their descendants and is now doing the same with millions of Asians, Latinos, and others.

It should be recognized, therefore, that while "ethnic" nations have usually possessed formal or informal means of assimilating non-ethnics into the nation, "civic" nations have usually mixed their political requirements for acceptance with cultural ones. The goal would seem to be to preserve a strong core national identity to which immigrants can adhere, while trying (as far as that is now possible) to prevent immigrant groups from creating closed and exclusive cultural communities of their own, on the assumption that a decision to move from one country to another implies a moral commitment to adopt a reasonable degree of adjustment to the host country's culture.[76]

National Citizenship

New forms of nationalism in Europe and the United States, if they develop, will do so over a period of decades and generations. At their heart, however, should be a very old idea: the supreme importance of national citizenship in terms of both rights and duties, and embodying "a love of country based on the common good."[77] In Western democracies, the rights may be greater than elsewhere, but the duties should be no less: duties to country, to society, and to fellow citizens.[78] Rights and duties are intimately connected, not only morally but practically—especially when it comes to taxation.[79]

As Yuval Levin has pointed out, we in the West take democratic citizenship far too much for granted:

> Our highly individualist, liberationist, ideal of liberty is possible because we presuppose the existence of a human being and citizen capable of handling a remarkably high degree of freedom and responsibility. We do not often enough reflect on how extraordinary it is that our society actually contains such people. A population of citizens generally capable of using their freedom well is the greatest achievement of modern civilization—greater even than the US Constitution and the market system, which depend upon such people.[80]

A central reason for the chronic malaise of Pakistan (and many other societies) is that it does not contain nearly enough of such people: citizens who understand themselves and the citizen body as a whole to possess by law and tradition rights under the state and duties toward the state—as opposed to individual and familial protections and benefits to be extracted from the state by whatever means available, and at whatever cost to everybody else.

The various attacks on the idea and content of citizenship in the West over the past generation have been damaging for every practical progressive cause, the struggle against climate change among them.[81] On one side of the assault we have the notion of "global citizenship"—praiseworthy if it is just a synonym for "a sense of commitment to the well-being of the planet" but absolutely vacuous and pernicious as a political principle, since it is divorced from every element of the real political content of citizenship.[82]

On the other side of the attack on citizenship we have the latter-day Marxian and anarchist notion of national citizenship as denoting "privilege" and therefore the oppression of non-citizens. Well, yes. These "privileges" include the right to vote and to stand for election; the right to participate freely in public debate; the right to enter the state services; the right to serve on juries; the right to equality under the law—in other words, together with associated duties, the entire relationship between the individual and the democratic state.

Of the two most influential documents on individual rights, the French National Assembly issued the Declaration of the Rights of Man *and the Citizen*" (my italics), and the US Bill of Rights guarantees the rights "of the people," in other words, the sovereign body of US citizens. The idea that a true citizen body can be replicated on a global scale is a delusion. The dissolution of national citizenship implies an amorphous mass of dissociated and powerless individuals, defenseless before the predations of a global overclass. Hardt, Negri, and Badiou may fantasize about leading such a mass in successful revolution. The rest of us might want to study the actual history of revolutions.

Central to the idea of citizenship has always been a commitment to serve and if necessary sacrifice one's life in the defense of the nation. This above all is what gives nationalism a moral element, which, if not morally superior to that of internationalism, certainly commands more emotional force:

> The political order seems less pure, less noble, and less human than the humanitarian order. It is not the universal human order, it is not addressed immediately to human beings as human beings. But precisely because of its nature, it weaves together efficiently the sentiment of the self and the sentiment of the other. Why? Because in the political order, the self and the other have something *in common*: precisely the political order, the body politic, the republic that is a *common thing*. . . . It is then possible for individuals to forget themselves and be willing to sacrifice themselves in a sacrifice that is both selfish and generous, the patriotic sacrifice.[83]

As climate change begins to bite, and global disorder grows, future nationalisms in the West and elsewhere will undoubtedly take on a strongly defensive character—not against other states, but against the menace of international chaos. At the moment, the idea of a "Fortress Europe" or a United States embodying liberal democracy is seen as a contradiction in terms. As the years go by, however, I believe that this will cease to be the case, and Western states will see themselves precisely as fortresses or redoubts of democracy in an increasingly troubled world.

As states elsewhere crumble or become more and more savagely authoritarian in order to survive, the need to preserve and defend our democracies will become more and more evident. This may well seem a bleak vision, but as pointed out in the introduction, there's not much sense in warning of climate change disaster and then suggesting that it is going to be nice and cuddly.

The result may be a new conception of democratic order, appropriate to a besieged city in which the population is dedicated to its defense. This may have certain analogies to wartime, in that it would emphasize the order as well as the democracy. The dedication to democracy may remain, but it would be conceived as preserving democratic civilization as a precious gift for future ages. As other states collapse and fall into ethnic civil war under the impact of climate change, in the West the need for inter-ethnic harmony and national unity may become more and more evident—assuming of course that Western states do not also collapse.

Together with grim objective circumstances, this could generate a new social ethic of social solidarity and shared duties of citizenship, which in turn could help to restore meaning to individual lives threatened by anomie. Given the horrors that will be occurring elsewhere in the world later in this century, differences of skin color among Americans and Europeans will come to seem insignificant. At some point, a reintroduction of national service could serve a number of purposes: the fostering of national unity and the overcoming of racial divisions; the rebuilding of a sense of public duty; the construction of defenses against climate change at home and abroad; the regeneration of local, national, and global environments; and ameliorating the consequences of the growth of automation and artificial intelligence.[84] The preservation of democratic civilization for humanity can therefore become a focus of common national loyalty and a basis for common national identity across lines of race and class.

4

Saving Capitalism from Itself

The capitalist of the future . . . will not be the ordinary dull rich man. He will either be a great criminal or a considerable patriot. If he is the first I hope that the law may be strong enough to keep him in bounds, but if he is the latter he may be a great ally of the state.

—John Buchan[1]

To tens of thousands of Englishmen engaged in daily toil, the call to "sacrifice" themselves for their country must seem an insult to their reason; for those conditions amid which they work make their lives already an unending sacrifice.

—Field Marshal Earl Roberts of Kandahar[2]

AN INCREASING NUMBER OF analysts are writing of the struggle of climate change as "the equivalent of war" and the need to put this effort "on a war footing." What they have lacked is any real sense of how modern populations are motivated to fight and win wars.[3]

On the other hand, writers like Nicholas Stern, Peter Newell, and John Mathews have rightly emphasized the great economic opportunities offered by the transformation of energy systems, and Stern has called this a chance of a new economic revolution comparable to those created by steam and electricity.[4] This is part of the inspiration behind Green New Deal thinking. But of course the great economic revolutions of the past, while they benefited certain classes and countries, had devastating effects on others. As the last chapter emphasized, even the countries that adapted successfully had to change themselves radically as a result, sometimes with very nasty social and cultural effects that took generations fully to reveal themselves.

This requirement of sacrifice for future generations does indeed link the struggle against climate change to war. It does not just take action against climate change quite out of the realm of normal politics under capitalism, whether of the democratic or authoritarian variety; it seems to go against the entire spirit and logic of modern consumerist liberal capitalism, with its relentless emphasis on short-term individual gratification.

Nationalism will also be vital if the effort to move to renewable energy and limit climate change fails and economies are forced into decline. It has been argued that this can be managed by states through organized "de-growth" (which, as Benjamin Kunkel has pointed out, would imply strict limits on both births and immigration).[5] In the longer term, this may indeed be inevitable. In the short to medium term, however, it is completely impossible as a political program, both in the (relatively) affluent West and in the developing countries that are chiefly responsible for the contemporary growth in emissions. Telling the impoverished masses of India that they have to abandon their hopes of prosperity and accept de-growth will be for a long time to come both politically and morally inconceivable. Even if Western countries were to adopt planned de-growth, India and China neither would nor could follow suit.

So if we are to have any chance of keeping climate change within reasonably safe limits, one version or another of a Green New Deal is the only way to go. If we fail, then economic decline will begin anyway, starting with South Asia—which is one of the strongest arguments to the policy elites in India and other developing countries for the need to move away from dependence on fossil fuels to drive growth.[6]

Fighting Climate Change through Capitalism

And yet, as the "eco-modernists" have emphasized and everyone but the radical environmentalists accepts in practice, timely action against climate change can only happen within and through a modified form of capitalism.[7] Any revolution in favor of a completely different system, with a different moral and cultural basis,[8] is only likely to occur *after* the present capitalist system has already collapsed because of climate change. The attempt to reshape economies to combat climate change is therefore better understood to be a continuation of the long struggle to save capitalism from itself, a struggle in which a key role was played not only by socialists but by nationalists.

A considerable and above all highly vocal part of the disparate camp of proponents of action against climate change, however, consists of committed opponents of capitalism tout court, like Naomi Klein, who declares that "climate change is a battle between capitalism and the planet." As far as the history of the past 250 years is concerned, it would be more accurate to say that climate change has been a battle between economic development and the planet, since every stage of this development since the invention of the steam engine has depended on the massive extraction and exploitation of fossil fuels—something that has been just as true for socialist as for capitalist economies.

For the future, what we can say is that climate change will be a battle between *uncontrolled* capitalism and the planet, just as over the past two centuries countries have experienced a long series of battles between uncontrolled capitalism and social and political cohesion. In some cases, a combination of sufficiently successful and flexible nation states and enlightened self-interest on the part of elites led to checks on capitalism (notably social welfare and public health and education systems) that preserved basic social peace and national unity. In others, the failure to achieve this led to revolutions of the left or right.

If existing capitalist states and societies continue to fail to take sufficient action to limit climate change, and the result is a catastrophe for humanity, then the case of the anti-capitalists will indeed have been conclusively proved, and capitalism will not just have collapsed but will have deserved to collapse. The problem is that capitalism will not somehow gently disappear. It will take billions of people, and civilization itself, down with it; and anyone who thinks that such a catastrophe will generate the sort of peaceful and just society of which the left dreams has really not been paying attention to 20th-century history.

Once again, this is not a threat that distinguishes between the different political systems of today, and the historians of the future (if there are any) are unlikely to distinguish much in their allocation of guilt if all fail. All of the major states of today, irrespective of specific political ideology, are creations of modernity and dependent on modernity. All share in the responsibility for its preservation. All have a vital interest in preventing the destruction of civilization.

Over the past 200 years, it has been proved, and beyond reasonable doubt, that capitalism is incapable of regulating and limiting itself. The nation state has to play a central role, based on the wider interests of the state and people. Unregulated financial speculation inevitably leads to crashes like those of 1929 and 2008. Even more important, without state and social controls, the capitalist search for increased profit tends inevitably to the immiseration of large parts of the population, the destruction of the environment, and the disintegration of society. Wise capitalists see this themselves, though it seems they have to learn the lesson over and over again.[9]

As Karl Polanyi wrote in 1944, "The idea of a self-adjusting market implied a stark utopia. Such an institution could not exist for any length of time without annihilating the human and natural substance of society; it would have physically destroyed man and turned his surroundings into a wilderness."[10] Eventually this leads to revolution or the weakening of the state to the point that it cannot defend the legal framework and maintain the national infrastructure that capitalism itself needs in order to work effectively. The difference today is that the threat is not to only one society but to human society as a whole through the generation of climate change.

It has also been demonstrated beyond doubt, from the examples of communist Russia, China, and elsewhere, that attempts to abolish capitalism are pointless and disastrous. The short-term human costs are horrendous, and in the long term these attempts are replaced sometimes by modified forms of state-led capitalism, as in China since 1979, and sometimes by the reintroduction of capitalism in its wildest and most destructive form, as in Russia in the 1990s.

The task therefore has been and remains to modify capitalism so as to make it reasonably responsive to the needs of societies as a whole—and in the context of climate change, of humanity as a whole. As already emphasized, reforming Western capitalism along the lines of a Green New Deal is also essential to rebuilding the national unity that will be essential if Western democracies are to survive the impact of climate change and other pressures in the decades to come.

Nationalism and Reform in Europe

Since the later 19th century, intelligent state elites have sought to prevent revolution and social collapse through social and economic reform. Where states have failed or been unable to do this, the result has all too often been catastrophe. With the memory of these catastrophes in mind, the consensus from 1945 to the 1980s supported "social market" capitalism, as it is called in Germany: a regulated capitalism (especially the financial system) and a fully funded welfare system and state health service funded by progressive taxation, strictly enforced.

From the 1870s to the 1940s, the establishment of social welfare systems was often closely linked to considerations of national security and national unity in the face of likely future conflict. This meant ensuring some kind of basic social security and health care for families.[11]

Industrialization and urbanization led to increasing military concerns about the physical fitness of the conscripts. This also helped inspire movements to reduce the appalling level of infant mortality in late 19th-century European cities.[12] In the United States, this impulse for reform took the shape of the Progressive movement, calls for "National Efficiency," and Theodore Roosevelt's "New Nationalism," which helped lay the basis for his cousin Franklin's New Deal.

Laws aimed at ending the ugliest aspects of industrial capitalism date back to the 1830s. The first systematic national insurance program was created by Bismarck in the 1880s, with the two joint goals of preventing revolution and fostering national unity.[13] As Bismarck told the German Reichstag, "[I have] lived long enough in France to know that the devotion of most Frenchmen to their government . . . was basically connected with the fact that most of them received pensions from the state."[14] A decade or so later, as Germany drew further and further ahead of Britain economically, the possibility of war with Germany became increasingly real for the British, and as social unrest and labor protests in Britain grew, sections of the British intellectual and political elites began to look to the German model.[15]

The "social imperialists" in Britain were a thoroughly eclectic bunch, drawn mainly from the imperialist wing of the Liberal Party, but also embracing Fabian socialists including the Webbs and (intermittently) George Bernard Shaw, "one nation" Conservatives, former colonial officials like Lord Milner and John Buchan, and the more farsighted sections of the military elites and their allies, like Field Marshal Roberts, Halford Mackinder, and Rudyard Kipling.

What brought them together in a loose alliance was belief in the defense of the British Empire, a conviction of the likelihood of a coming world war in which national unity would be tested to the limit, a certain professional middle-class contempt for the hereditary aristocrats and professional politicians who dominated British governments, and a deep fear of revolution, class warfare, and social disintegration.

At the core of this thinking was also a belief in "national efficiency": that the British state needed to be thoroughly reformed and given increased powers, including to shape and guide the economy.[16] H. G. Wells called it "the revolt of the competent." Or according to Winston Churchill (then a member of the Liberal government), "Germany is organised not only for war but for peace. We are organised for nothing except party politics."[17] The social imperialists' vision extended beyond social insurance to urban planning, public health, and educational reform.[18]

Indeed, in all the Western European countries, the success of their different programs of social imperialism before 1914 would be reflected in the extraordinary endurance and self-sacrifice of their armies; in comparison, the Russian imperial state's inability to provide minimal social welfare to its citizens contributed greatly to the collapse first of the army and then of the state in 1917.

The social imperialists generally believed in the need for a new guided "national economy," the need for higher progressive taxation to pay for both social reform and military preparation, and in limits on free trade to protect British industries and imperial economic unity ("imperial preference"). They were therefore in rebellion against the free-market orthodoxy that had dominated both political parties since the repeal of the Corn Laws almost sixty years earlier. In an interesting parallel to the present, their thought developed in the context of the decline of British industry in the face of growing international competition and the steep growth in relative importance of the City of London and financial services.

The social imperialist ideology in Britain was a politically and morally ambiguous one. On the one hand, there were definite parallels with the European tendencies that contributed to fascism and German National Socialism after the First World War. On the other hand, through their overlap with the traditions both of the British Labour Party and of "One Nation" Conservatism, the social imperialists contributed to the growing national consensus that eventually created the British welfare state after 1945; just as in Germany, Italy, and France, the social democratic tradition joined with Catholic social reformism and conservative reform from above to produce the "social market economy."[19]

In Britain, social imperialism—though under new names—was strengthened and eventually triumphed as a result of the wars, and especially the Second World War, in which the Conservatives and Labour worked together in government.[20] In the course of that war, Labour became deeply patriotic and the Conservatives accepted rigorous state guidance of the economy. The Beveridge Report of 1942, which laid the basis for the post-1945 welfare state, was a product of the Second World War; and Beveridge's own thought was originally rooted in social imperialism.[21]

It can therefore be said that rather than serving as an analogue of German National Socialism, the British social imperial tradition in fact played a key part in saving Britain from fascism. The great bulk of the tradition flowed into moderate socialism and one-nation Toryism.

So just as the social democratic and progressive movements in Europe and the United States turned out to have as their result saving capitalism from itself, so social imperialism in Britain could be regarded as saving nationalism from itself. Major Clement Attlee, Military Cross, and Captain Pierre Mendes-France, Croix de Guerre, were not after all the same as Corporal Adolf Hitler, Iron Cross.[22]

Bernard Semmel's standard work on social imperialism was published in 1960, the year that I was born; and it makes both depressing and inspiring reading today. Semmel takes it for granted that as a result of the wars and the welfare state, unconstrained free-market capitalism has been ended forever; but also that the British Labour Party has abandoned forever dreams of replacing the nation state with an international proletarian order: "Today, the Cobdenites [radical laissez-faire liberals] and the international socialists are virtually extinct breeds."[23]

Sixty years on, unconstrained free-market capitalism has once again been running amok for a generation, with disastrous results; and many socialists in the West have once again abandoned loyalty to their nations in favor of a return to fantasies of a borderless progressive world guided by themselves—and looming in the background, unaddressed, is the threat of climate change to all existing states. The task then is to develop a new version of social imperialism without the imperialism, racism, eugenics, and militarism. And to those who say that this is a contradiction in terms, I would say: it has been done before. Keynes defended himself in 1940: "I have been charged with attempting to apply totalitarian methods to a free community. No criticism could be more misdirected. In a totalitarian state the problem of the distribution of sacrifice does not exist. . . . It is only in a free community that the task of government is complicated by the claims of social justice. . . . The aim of these pages is, therefore, to devise a means of adapting the distributive system of a free community to the limitations of war."[24]

Studying the British Home Front in the Second World War with my daughter, I was struck by how many of the appeals find their echo in environmentalist thinking today, from recycling to conserving energy, fuel, and water: "Is Your Journey Really Necessary?"; "Walk Short Distances!"; "Have You REALLY Tried to Save Gas? Join a Car Club!" (United States); "Save All Paper and Cardboard for Salvage"; "Save All Cans!"; "Grow Your Own Food!"; "War Time Cookery: To Save Fuel and Keep Healthy"; "Take a Five Inch Bath!" People grumbled—I expect I shall grumble myself in the future if I live long enough to see these measures implemented again—but they generally obeyed, for the sake of the common war effort.

In both Europe and the United States (at least when it comes to Social Security and Medicare), the achievements of social reform in the first half of the 20th century have left behind in the consciousness of the people a sort of "moral economy" (in a sense very different from that meant when Bill Gates uses the term), a belief that the legitimacy of the state depends on its providing certain benefits to the population as a whole.[25]

In Russia, collapse of this contract between state and society in the 1990s contributed greatly to the widespread feeling that people have no duties to the state and that corruption is therefore in some sense morally justified.[26] This is what threatens us in the West too if the rich are visibly able to disobey the laws that apply to the rest of the population, especially when it comes to paying taxes.

The "New Nationalism" in the United States

In the United States, the impulses that in Britain produced social imperialism fed into the progressive movement and Theodore Roosevelt's and Herbert Croly's concept of "the New Nationalism"—though with a relatively smaller focus on welfare and a larger one on regulation of capitalism and national efficiency.[27] The situation in the United States that produced this tendency had certain analogies to the present as well as important differences. It followed a period of great economic growth, although the fruits of this growth had been very unevenly distributed. By 1896, it was estimated that 1 percent of the population owned over half of the wealth in the United States, and that 12 percent owned nine tenths.[28]

As in the United States today, the enormous concentration of wealth and monopolization of key industries in the United States at the start of the 20th century threatened some of the founding myths of the American republic: the idea of a basically middle-class (or in Jefferson's older formulation, "yeoman farmer") society with a rough equality of conditions (something that Tocqueville thought an essential basis of American democracy), and equality of opportunity for all citizens in a free economy.

Corruption had always been part of US public life; but the new huge concentration of wealth meant a huge concentration of political power in the hands of "robber barons" like John Pierpont Morgan and Andrew Carnegie—just as today, huge political power is concentrated in the hands of super-rich individuals and companies whom the Supreme Court has allowed virtually unlimited ability to fund politicians and campaigns.[29]

This power helped the great corporations to form "trusts" aimed at creating monopolies in particular sectors, while the railroads used their domination to impose grossly unequal tariffs between different regions. Out of this came the image of the "Octopus" (the title of Frank Norris's novel of 1901 about the struggle between California farmers and the Pacific and Southwestern Railway), a monster whose tentacles stretched into every part of politics and the economy.[30]

Simultaneously—as in recent decades—there took place massive immigration to the United States from Europe, involving groups who had not previously been present in the USA, did not speak English, and mostly belonged to religions (Catholicism and Judaism) alien to the old Anglo-Scottish Protestant tradition. This caused deep anxiety in the older population. The enormous growth of American cities led to worry at the increasing role in politics of urban political machines run by immigrants ("Tammany Hall"), which merged with anger at their corruption, outrage at the dreadful conditions in the urban slums, and fear of urban revolt, of epidemic disease, and of disasters like the Chicago and Boston fires. This was accompanied by a combination of social concern and moral outrage at the vice and addiction that flourished among the new urban poor—the addiction in those days was to alcohol but was no less of a scourge for that.

In essence, the American elites in the early 20th century—Republican, Democrat, and independent—had to learn something that Republican hardliners have spent the last generation laboriously forgetting again: that you cannot run a great modern mass society and economy on the basis that the answers to every question are to be found in a sacralized 18th-century constitution and blind faith in unrestrained free-market economics; and that many areas of modern government need to be managed by technocratic experts, not amateur political cronies appointed through the spoils system. In the analysis of the leading Progressive thinker Herbert Croly in 1909: "The experience of the last generation plainly shows that the American economic and social system cannot be allowed to take care of itself, and that the automatic harmony of the individual and the public interest, which is the essence of the Jeffersonian democratic creed, has proved to be an illusion."[31]

Like the British social imperialists, but very unlike most social reformers of today, Croly's work was also profoundly nationalist, dedicated to the American national interest and to instilling in the American state and population—and especially the new immigrants and their children—a new sense of national purpose. And while Croly's grander hopes were not fulfilled, the successful integration of the immigrants (in part through a vastly expanded state education system that the Progressives had championed) helped create the national consensus that a generation later supported the New Deal, and the "trust-busting" measures of Theodore Roosevelt and Woodrow Wilson greatly reduced monopolization.[32]

Croly's work formed the basis for Theodore Roosevelt's "New Nationalism": the program of the short-lived Progressive Party, with which Roosevelt attempted to regain the presidency in 1912, and which he set out in a famous speech of 1910 in Osawatomie, Kansas. Roosevelt's advocacy of social and political reform was, of course, underpinned by an ardent and convincing personal commitment to nationalism.

Roosevelt's platform included points that are of great relevance today, including attacks on the power of special interests and monopolies, and a demand that business executives should be held personally responsible for the crimes of their corporations. Roosevelt called for the restoration of what he called the "square deal": the principle that in America, hard, honest work was adequately rewarded, that every hardworking American had the opportunity to get ahead, and that the equality of the vote should not be subverted by the rich: "The man who wrongly holds that every human right is secondary to his profit must now give way to the advocate of human welfare, who rightly maintains that every man holds his property subject to the general right of the community to regulate its use to whatever degree the public welfare may require it."[33]

Roosevelt can also be called the political father of environmentalism in the United States. The founder of the national park system and preserver of tens of thousands of square miles of wilderness, he also worried deeply about the wasteful exploitation of America's natural resources:

> National efficiency has many factors. It is a necessary result of the principle of conservation widely applied. In the end, it will determine our failure or success as a nation. . . . Conservation means development as much as it does protection. I recognize the right and duty of this generation to develop and use the natural resources of

> our land; but I do not recognize the right to waste them, or to rob, by wasteful use, the generations that come after us. I ask nothing of the nation except that it so behave as each farmer here behaves with reference to his own children. That farmer is a poor creature who skins the land and leaves it worthless to his children.[34]

In 2011, President Barack Obama spoke at Osawatomie, referencing Roosevelt's reformist program, reminding people that this was a former *Republican* president, and comparing the illegitimate power of the rich then and now.[35] Obama's ability to draw on bipartisan American national traditions were part of the reason for his being twice elected the first black president of the United States.

In his speech, however, Obama did not really evoke the spirit of nationalism, of national efficiency and competition with other nations that permeated Roosevelt's attitudes. Moreover, by 2011 Obama had already lost by far his best chance to implement radical reforms in the immediate wake of the financial crash three years earlier. Instead, he allowed himself to be ingested by the same financial establishment that has dominated economic policymaking for decades, under Democratic as well as Republican administrations.[36] A future US president who hopes to push through a Green New Deal will need to combine Obama's appeal to core US traditions with much greater radicalism, backed by a much stronger appeal to nationalism.

State Reform and Action on Climate Change

As Obama emphasized, the first target of any reformist program today must be what Roosevelt called "lawbreakers of great wealth," especially in the context of tax avoidance. The first need of states in pursuing reforms, including the policies needed to limit climate change, is to raise the money to pay for them.[37] The mass motivation and state strength that have been central themes of this book are in part a means to this end. Globalization, deregulation, and the power of the global overclass mean that even very powerful states once again face an ancient challenge in this regard: "The West was faced with the paradox of immensely wealthy individuals and an extremely poor treasury. . . . For such individuals, their devotion to *Romanitas* meant primarily fidelity to an elite cultural tradition and the preservation of the immunities and privileges of their class."[38]

The United States and the EU in the first quarter of the 21st century AD? No, the Western Roman Empire in the last quarter of the fourth century AD. Today, you can probably scratch the cultural tradition bit as far as the US wealthy are concerned, but the identification of American democracy with their own privileges is as strong as it ever was.

The great Roman senatorial landowners had used their power largely to emancipate themselves from paying taxes, thereby passing the burden on to the mass of the population. So crushing did this burden become that it has been suggested that by the fifth century, many Roman citizens in the West actually preferred to be conquered by the barbarians.

The early modern state was shaped through the struggle with such "overmighty subjects"—who, however, never went away and are now back with a vengeance, though in a new form. Modern nationalism itself was called into being largely by the establishment of sovereign peoples without legal distinctions of hereditary class.

Globalization and financial deregulation have greatly increased the opportunities for tax avoidance by elites all over the world.[39] As I witnessed in Russia and Ukraine after the Soviet collapse, open borders, free-market reforms, and privatization became a license for predatory elites (comprising both old and new elements) to plunder state property and transfer the money to the West and various tax havens, with the enthusiastic help of Western banks. This was then euphemized by the Western media as "capital flight."

Under cover of "democracy," these elites instituted a form of kleptocratic oligarchy (tamed but also continued in a modified nationalist form under President Vladimir Putin), which seemed to be pointing Russia down the road toward the condition of Nigeria. This was the reality of the liberal internationalist dream of weakened nation states in a globalized world.[40]

And as Franklin Foer wrote in the *Atlantic* magazine (quoting the warnings at the time of the courageous and farsighted CIA station chief in Moscow, Richard Palmer), the economic, political, and moral disaster has not been Russia's alone. As a result of this colossal crime and the complicity of Western financial institutions, "the values of the kleptocrats became America's own"—and the city of London's, I might add. Foer describes "the collective shrug in response to tax avoidance by the rich and by large corporations, the yawn that now greets the millions in dark money spent by invisible billionaires to influence elections. In other words, the United States has legitimized a political economy of shadows, and it has done so right in step with a global boom in people hoping to escape into the shadows."[41]

Russia became only one piece in a global pattern of tax evasion and money laundering on the part of rich individuals and corporations. This has indeed weakened the power of nation states: their power to fund hospitals, schools, pensions, and infrastructure, and—most important of all for the future—their ability to raise the massive amounts of money that are needed to reshape economies and societies in order to limit climate change.[42]

In the United States, the wealthy manage to escape most taxes—perfectly legally—by using their power to persuade the US Congress to pass tax cuts exempting them; and quite apart from this, around one sixth of the taxes owed are never paid, partly because of tax havens.[43] In Pakistan, the elites (and much of the rest of the population as well, to be fair) simply laugh at the tax collectors.

In China, the ferocious anti-corruption campaign launched by Xi Jinping is in essence a new stage in the age-old struggle of the Chinese state to prevent state elites from escaping state control and state taxes; for repeatedly in the past, elite corruption, loss of revenue, and consequent inability to raise and feed the masses of workers necessary to maintain the flood control systems has cost dynasties the Mandate of Heaven.

Both technically and financially, the move to a "Green Capitalist Economy" is already possible in the West, and even in China, despite far smaller wealth per capita (creating the financial resources is a great deal more difficult in India and Africa).[44] Alternative energy is rapidly becoming both cheaper and more efficient. Nuclear energy is available at small risk (though very large costs of installation). Nuclear fusion may make a much safer and cleaner nuclear option possible in the future. Widespread and cheap public transport can replace much automobile traffic. Given the colossal scale of global financial markets, a relatively small tax on financial transactions, rigorously enforced, would provide a large part of the revenue.[45]

For that matter, the more than $1.5 trillion spent on bank bailouts in 2008–2009, the enormous sums put into directionless "quantitative easing" in the years that followed, and the almost $6 trillion that the United States has spent on the wars in the Middle East and Afghanistan all indicate just how much money could be made available if the political will were there.[46]

Given the experience of attempted reforms since 2008, however, it is difficult to exaggerate the power and determination of opposition to reform from the banking and energy sectors.[47] It is melancholy to recall in this connection that Franklin Delano Roosevelt (FDR), now generally credited with saving American capitalism in the 1930s, was called a "communist" by many capitalists at the time and had to conduct a bitter fight with a majority of the US Supreme Court. And that was when the official unemployment rate in the country stood at almost 25 percent. Moreover, the capitalist opponents of a green economy today can also count not just on masses of people deluded by propaganda but also on many workers and regions who really will be badly affected by the move away from coal and oil.

As France has demonstrated, even with a highly efficient public transport network, many motorists will fight tooth and nail to prevent small restrictions on their driving. Given the relatively slow improvement in the efficiency of hybrid cars, tremendous state influence including much higher fuel prices will be needed to move a majority of motorists over to them within the next two decades. Governments committed to action against climate change will also have to be tough with their own Green supporters clamoring for the abandonment of nuclear energy; and they will need to insist on funding attempts to develop fusion power, which may promise immense supplies of nuclear energy without the risks and the nuclear waste of existing nuclear power.[48]

Ruthlessness along Chinese lines is neither desirable nor practicable in the West (though if things are allowed to continue as they have over the past generation, that day will come). Nonetheless, tough action to raise revenue from the elites has to be an essential foundation of any Green New Deal, both for the sake of the money itself and to convince ordinary people that the call for sacrifices is legitimate and the elites are not shuffling the sacrifices onto them—the impression disastrously created by President Macron in France in his fuel tax.

The measures necessary include those introduced and advocated by Theodore Roosevelt, Woodrow Wilson, and FDR plus some new ones: the return of tough progressive taxation; the closure of tax loopholes, including the abolition of the (morally empty) distinction between tax avoidance and tax evasion; international action (above all by the United States and Britain) to eliminate tax havens and centers for money laundering; much tougher punishments for tax evasion, including exemplary sentences for prominent figures, *pour decourager les autres*; a return to the spirit of the strict regulation of banks that existed from the 1930s to the 1980s and the nationalization of particularly egregious offenders as a warning to the others; and a shift from fining corporations for crimes to fining the executives responsible.

Some of these proposals have been incorporated in the platforms of Elizabeth Warren and other Democratic candidates in 2020.[49] It would be disastrous, however, if the response to global warming remains overwhelmingly a liberal and socialist initiative. As the emergency grows, security establishments need to override the interests and demands of specific economic groups in society and make this a central concern of the state as a whole. Given the growing effects of climate change, this shift is indeed bound to come sooner or later. The question is whether it will come soon enough to prevent catastrophe.

National Efficiency

As the example of the 1950s and 1960s in the West and many developments in East Asia demonstrate, there is nothing inherently incompatible between such measures and successful capitalism. This is also true of the purpose of this taxation, which in the end boils down to building new infrastructure—a state task since ancient times. The critical role of the state in building the infrastructure on which capitalism depends has been obscured in the West by the hegemony of the British and American examples, which did not require state investment to build the canals and railways: in the British case because British imperial trade, the plundering of India, and slave plantations created the necessary private financial surpluses; in the United States because the railroads could be financed by enormous amounts of new land taken from the Native Americans and allotted to the railroad companies.

Across the rest of Europe and Asia, however, the state played a critical role in building railways, either for directly military purposes (as with the British railways in India) or to create industrial economies capable of supporting modern militaries and sustaining economic competition with rivals. Thus in the 20 years before 1914, spending on railways came second only to spending on the military in the state budgets of the German Empire. In the United States from the 1940s to the 1970s, the role of the state and the military in helping to drive technological innovation was almost universally acknowledged.[50]

California in some respects is a model for state action against climate change through capitalism, and based on a consensus of a sufficiently broad spectrum of the political elites. Despite the obstruction of the Obama administration's federal efforts to reduce emissions by a Republican-controlled Congress, California, remarkably, managed to hit its 2020 target for reductions in emissions four years early, in 2016. This was thanks above all to agreement between enough of the state's Republicans, led by the "Governator," Arnold Schwarzenegger, and the Democrats, led by Schwarzenegger's successor as governor, Jerry Brown. As this book has emphasized, this kind of consistency is essential if effective action against climate change is to be maintained.[51]

But California also illustrates the constraints of inadequate taxation and trying to rely chiefly on private finance obsessed with short-term gain. If California were an independent country, it would be the fifth largest economy in the world. It is home to some of the world's greatest high-tech industries. Yet despite almost a generation of talk and a consensus on climate change between Republican and Democratic governors, California cannot even raise the money to pay for a high-speed rail link between the Bay Area and Los Angeles (of the kind which now link even medium-sized Chinese cities). Why not? Two reasons are the refusal of US federal administrations to provide funding from overstretched federal resources; and the unwillingness of US banks to invest in anything that will only make a profit in the long term; but another is Proposition 13 of 1978, whereby in a referendum Californians voted almost two to one for strict limits on taxation. If ordinary Californians are to be persuaded to lift these limits as part of the struggle against climate change, they must see benefits for the mass of the population, and they must see corporations and the wealthy paying their fair share in taxes.

Nicholas Stern has written, "The scale of emissions reductions associated with avoiding grave risks of climate change implies nothing short of a new energy-industrial revolution. Experience of past technological revolutions suggests that they are associated with waves of growth and prosperity of two to three decades or more."[52] I must confess that I am a little skeptical of this. As Lord Stern himself has famously written, to date, "Climate change is the greatest market failure the world has ever seen," and it will take a very great deal of state intervention and political will to change that. There are indeed great economic opportunities in the change to technologies of alternative energy and energy conservation, but they do not in themselves seem likely to rival the impact of the steam, electrical, automobile, or aircraft revolutions, all of which represented vast increases in technological and productive capacity and attracted huge amounts of private investment. These were capitalist revolutions, enabled by states but also fully motivated by profits (allowing, of course, for crashes along the way).

In the shift to alternative energy and energy conservation, by contrast, the capitalist profit motive is very much weaker and the need for state intervention much stronger. Alternative energy would in the long run save the capitalist system; but in the dreadfully short horizons of contemporary capitalist profit-driven vision (and especially its Anglo-American variant), it does not offer anything like the same incentives.

Relying on market calculations and the steadily growing efficiency and affordability of alternative energy and electric cars, the British economic analyst Kingsmill Bond has predicted that demand for oil will peak in the 2020s, that renewables will overtake fossil fuels around 2050, and that by the end of the century renewables will account for the overwhelming majority of energy generation.[53] For an investor, this seems to make good sense; but as the scientists are telling us, we need to bring fossil fuel use to far below 50 percent of energy generation by 2050 if we want to avoid really dangerous rises in temperature. So while state ownership (except of certain large banks) is not often desirable or necessary, a strong degree of state direction and control most certainly is.[54]

The predictions of writers like Deirdre McCloskey, who believe that free-market capitalism alone is capable of adopting adequate measures against climate change, are all too obviously not coming true. It has been a generation or more since arguments along these lines first started appearing, and in that time target after target for reduction of emissions has been missed even as the task of meeting them in the necessary timeframe gets more and more difficult, while powerful capitalist interests have resisted any serious action at all.[55] Moreover, such writers almost completely fail to understand the radical difference between the limited effects of capitalist growth on the world in previous centuries and the vastly increased impact in the contemporary era—if only because of the difference between one billion people in 1800 CE and 7.7 billion in 2019.

Alternative energy is getting better all the time; but it still only replaces existing electricity generation, which is fine at doing what it was meant to do: producing cheap electricity. As David Wallace-Wells grimly remarks, after improvements in cost and efficiency that greatly exceed the most optimistic visions of a generation ago, "We are . . . billions of dollars and thousands of dramatic breakthroughs later, precisely where we started when hippies were affixing solar panels to their geodesic domes. That is because the market has not responded to these developments by seamlessly retiring dirty energy sources and replacing them with clean ones. It has responded by simply adding the new capacity to the same system."[56] Thus India has made remarkable progress in recent years in the introduction of renewable energy. By 2019, 35 percent of electricity generation came from this source, and India was four years ahead of schedule in the introduction of solar energy. This has however mostly covered the growth in overall electricity generation that has accompanied India's economic growth. Few existing coal-fired plants have been shut down, and as of 2019, 55 percent of Indian electricity was still generated by coal.[57]

It is true that when installed, alternative energy and energy-conserving processes can be run at much lower cost because they do not need fuel, but from a purely free-market point of view this does not justify the cost of replacing existing capacity: the 1–3 percent of GDP per year that Lord Stern and others estimate will need to be invested in alternative energy if electricity generation is to be virtually carbon-free by 2050.[58] Initially at least, a large part of this will have to come from states; and heavy taxation of carbon is also essential if industries are to be moved toward alternative energy. This will create many new jobs, but it will also destroy many old ones—at a time when, as Chapter 2 emphasizes, huge numbers of jobs will also be at risk from automation and artificial intelligence.

If developing countries are not to wreck limits on climate change by burning increasing amounts of coal and oil to support their development (and thereby in the end negate advances by developed countries), then the West, China, and Japan will have to bring both enormously increased aid and a great deal of pressure to bear. The aid should go to the installation of alternative energy and nuclear power and can be justified to electorates if it is explicitly linked to the purchase of this technology from the donor countries. China appears to be heading toward something like this in parts of its investment program in Pakistan. The pressure should take the form of prohibitive tariffs against countries and companies that do not meet strict emissions standards.

In the United States and Europe, investment in alternative energy fell severely between 2011 and 2013 as a result of changes in government policy. Only China continued to grow (by 2018, China accounted for $126 billion out of $279 billion invested in renewable energy worldwide).[59] Assuming that the Chinese Communist Party sticks to this line and remains in power, we can expect strong growth of renewables to continue for generations to come.[60]

Hence the repeated emphasis in this book on the need for Western democracies to develop strong cross-party consensuses on climate change, based on perceptions of threats to national security. At present, the contrast between China's record and that of much of the West should make painful reading for any democrat—especially given China's far lower levels of per capita GDP; and in China over the past two generations, nationalism has in effect replaced communism as the legitimizing ideology of the state.[61]

The achievements of China compared to the United States (less so Europe) also apply to public transport. As the Green New Deal proposals have emphasized, an immense expansion of public transport will be essential in the United States. So too will be the construction of new energy-efficient housing, which, to make a serious impact, would require the replacement of great quantities of existing housing. That cannot be profitable unless energy prices rise far above their existing levels. Quite apart from climate change, the benefits to societies from the move to cleaner, less polluting technologies will be immense.

In terms of present "known unknowns," what would both vastly help the fight against climate change and represent an equivalent of previous economic revolutions would be if someone could invent a way to mass produce that miraculous substance called Graphene (a laboratory-generated form of carbon that is around one hundred times harder than steel but so thin and light as to be transparent, and which is also an extremely efficient conductor of heat and electricity); but we still have no idea whether that will ever be possible.

In the meantime, there seems no alternative to massive state involvement, in terms of direct investment, subsidies, incentives, and compulsions; and this needs to happen very quickly if disaster is to be avoided. Hence the talk of placing this effort on a "war footing," and hence the argument of this book that as in war, the sacrifices and austerity required will need to be motivated by nationalism and softened and legitimized by state-led social solidarity. The production of alternative energy technology can certainly help sustain developed economies, but it seems unlikely that it can return them to the levels of growth that sustained the social market economies from the 1940s to the 1970s.

Rather than aiming at short-term growth and then trying to tack reduction of emissions onto it, governments will have to extend their time horizon and look at the savings to economies over the next generation at least. At present, that is not something that contemporary Western capitalism seems capable of; but as the example of "social imperialism," the "new nationalism," and their legacies demonstrate, we should not despair of this happening over time.

Paul Collier has pointed out that American and British firms were not always so obsessed with short-term shareholder "value" as they are today.[62] Under the impact of a growing crisis and as part of a wider transformation of societies and cultures, they may shift back to a long-term view and a sense of social responsibility. Unfortunately though, with global warming breathing down our necks, we cannot wait the decades necessary for such a transformation to happen. States will have to take the lead, and do so quickly.[63]

Given predictions that if unchecked, the growth of air travel by 2050 may increase emissions from aircraft by 300 percent compared to 2019 figures, then either airlines will have to move to alternative fuels on a scale that may not be practicable, or it will be necessary to move to much slower forms of air transport, or greatly to reduce air travel by rationing or taxation. Speaking for myself, I wouldn't mind the first outcome at all, but I would absolutely hate the second. I will accept it as a necessary part of the struggle against climate change, but only if the state demands it of the population as a whole, irrespective of income. Indeed, in countless areas of life, some degree of rationing together with the attendant reshaping of economic expectation and freedom will be essential in the struggle against climate change, pollution, ecological degradation, and resource depletion. I hope that green technology will improve so much and so quickly that this will not be necessary—but I wouldn't count on it.

States also need to pursue research into geo-engineering solutions, despite fundamentalist ecological objections. Given the scale and urgency of the threat from climate change, it would be grossly irresponsible not to investigate these possibilities. Equally, however, it would be grossly irresponsible to delay actions radically to reduce emissions in the hope that possibly, at some point in future, technology may come along to make this unnecessary. Furthermore, even if the technology were available, attempts to manipulate the earth's atmosphere as a whole look desperately risky.

Limited efforts by groups of nations to cool parts of the Arctic may be possible at some stage (for example, by spraying clouds with seawater). If so, they should definitely be attempted, given that the main potential tipping points for runaway climate change are situated in the Arctic. The British government is creating a new center in Cambridge to investigate the development of technology to do this.[64]

One possibility of mitigation involves new technology to capture CO_2 at the point of emission from power stations and factories. This should be concentrated on coal-burning plants, since coal is the worst fossil fuel of all for carbon gas emissions and still generates 38 percent of global electrical power—and in terms of absolute consumption has doubled since the year 2000. This proportion has not changed since the time of the Kyoto Agreement in the 1990s, above all because of the huge increase in Chinese power generation over the past decades.

Once again, though, the scale required is enormous. To reduce emissions by 20 percent over the next 30 years would require the construction of some 3,000 large carbon capture plants, from a mere handful today—an effort and expense comparable to that involved in switching to renewable energy, and one that could only be accomplished by state action. So the free-market ideologues who argue that geo-engineering is a way of limiting climate change without reforming capitalism are fooling themselves and everybody else.[65]

The Chinese state's calculation concerning coal seems to be a waiting game; rather than the Chinese want to see whether new technologies, either to replace coal production or to trap CO_2, appear in the 2030s and '40s. Economically speaking, this does indeed make sense; but in the intervening years, we may have generated so much more CO_2 that climate disaster will already be likely even if emissions decline radically after that. Much of the carbon dioxide generated today will remain in the atmosphere for thousands of years. So while we obviously should continue an intensive search for new energy technologies, we need to push ahead quickly now with the alternative technologies that we already have.[66]

At the heart of China's relative success in moves to limit emissions and introduce alternative energy has been consistency and predictability, stemming from the commitment of the Chinese government to the science of climate change, and the fact that the Chinese government is not going to be voted out of office at the next elections. This allows long-term planning and investment and the incremental introduction of radical changes over a decade or so. This long-term consistency, as well as state dictation, is critical to the ability of Chinese businesses to change and adapt.

The same has been true of the water conservation policies of Israel, a state that virtually embodies the principle of "national efficiency." As Chapter 1 indicated, water shortages will be a critical issue for much of humanity even before the effects of climate change really kick in. Israel has led the world in water conservation, through pioneering achievements in drip irrigation, self-powered desalination plants using reverse-osmosis, and waste-water recycling. As a result, Israel has been spared the effects of the droughts that have plagued the Middle East over the past generation; and pride in "making the desert bloom" is a key part of Israeli national identity.[67]

In Israel, 85 percent of purified sewage is recycled—more than three times the rate in Spain, the next country on this scale, and around eight times the proportion in the United States. As mentioned in Chapter 1, however, the most important factor of all has been getting the population to save water by making them pay high prices for it. This was only politically possible because of an awareness rigorously instilled in the Israeli population: "Every drop counts. Every Israeli you meet has had it drummed into them that faucets shouldn't be left on. Water conservation has been a part of elementary school education in Israel for generations. . . . It's just a part of the culture."[68]

This consciousness does not exist in isolation but is part of a deep sense of nationalism and national insecurity. The Israeli approach to water also contradicts the argument of Bjorn Lomborg and others that it would be better to wait to introduce measures against climate change because future generations will be richer and more able to pay for them. Israel is of course a rich country, but it was a lot poorer 60 years ago when it laid the foundations of its water conservation strategy. And had it not done so, it would be a poorer country today, with a lower quality of life and lower national security. Singapore, another country with a strong sense of national vulnerability, has also sought to make water efficiency a key part of national identity and national pride.

The challenge for Western politicians aiming at effective strategies against climate change is to create by democratic means the sort of national consensus that will make such consistent long-term strategies possible.[69] The first New Deal, which saved American capitalism from itself, and went on to save Western democracy in the Second World War, was founded on two things: a recognition that capitalism had fallen into deep crisis, and the creation of a new long-term national consensus.[70]

If we want to limit climate change and to survive its consequences, this is the sort of national spirit that every nation will need to cultivate.

5

The Green New Deal and National Solidarity

When there are no shared goals or vision of the public good, is the social contract any longer possible?

—Allan Bloom[1]

The issue of environmental quality is one which transcends traditional political boundaries. It is a cause which can attract, and very sincerely, liberals, conservatives, radicals, reactionaries, freaks, and middle-class straights.

—Russell Kirk[2]

THE CENTRAL GOAL OF the Green New Deal must be to devise strategies that limit anthropogenic climate change. Measures to increase social solidarity are an essential accompaniment to these strategies, to build solid national majorities behind action, to allow governments to ask for the sacrifices necessary in the struggle against climate change, and to strengthen the unity and resilience that societies will need to withstand the effects of climate change and other shocks over the next century. All other policies, however good in themselves, must be judged by their contribution to these two objectives. Without something like Green New Deals, sustained over many years, Western liberal democracies won't last long enough to be overwhelmed by the direct effects of climate change.

Progressives need to keep firmly in mind that if we fail to limit climate change, the resulting world is extremely unlikely to be friendly to the causes they have tried to load onto the climate change bandwagon.

The battle against climate change is *in itself* a battle to defend a progressive and liberal civilization. It does not need extraneous causes loaded onto it. Above all, it does not need causes that will make the creation of national majorities behind action against climate change even more difficult.

The Greens in Europe

On reading the electoral program of the French Green Party (Europe Écologie-Les Verts), my first thought was that no serious person could take it seriously.[3] My second thought was that things were far worse than that. There is a very powerful force in France that takes such programs seriously because they get votes from them. Unfortunately, that force is the National Front.

The French Greens' "profession of faith" begins: "We have the right to happiness"—a mixture of woolly Rousseauist narcissism and irresponsible hedonism if ever there was one. If I believed that, I wouldn't get my philosophy from the Greens; I'd get it from W. C. Fields: "Ah yes, a man has to believe in something—and I believe I'll have another beer." The statement then immediately qualifies this by declaring that "We have the right to want the best for those we love." Well that's OK then. I'll worry about my family and a handful of very close friends, and everybody else can go hang. How many people can I genuinely "love," for God's sake? I certainly won't worry about my great-grandchildren in the year 2100 CE. Why should I? I'm never going to meet them.[4]

This sort of thing can be dismissed as empty sermonizing, but other parts of their program are deadly serious—or would be if they were ever implemented. The summary of their "plan of action" for the European elections of May 2019 reads as follows:

> Let us construct [Europe] around the goal of escaping from carbon and nuclear [power] and of investing massively in ecological transition. . . . With the help of a reformed and unified asylum policy, let us construct a Europe capable of dealing with the question of welcoming the men and women whom global tumult has thrown onto the roads at the risk of their lives, a Europe capable of ensuring peace on the continent.[5]

It would be very hard to exaggerate the irresponsibility of this statement. By equating fossil fuels with nuclear energy, the first part is in effect deeply hostile to real French action against climate change. France's relatively low rate of carbon gas emissions is due above all to the fact that nuclear energy accounts for around 80 percent of its electricity generation—and has never suffered a single significant accident. By trying simultaneously to abandon both nuclear and carbon energy, this "action plan" if implemented would either plunge France into darkness or ensure that the stated goal of eliminating carbon fuels could not possibly be met.[6]

The second part of the statement calls for more acceptance of asylum seekers. Since elsewhere in the party manifesto there is a call for the distinction between asylum seekers and economic migrants to be abolished, this is in effect a call for open borders. Such a program would tear France apart internally as well as tearing the European Union apart along regional lines. It might then bring the EU back together again through a universal triumph of populist chauvinism. This would admittedly after a fashion ensure the third part of the statement, peace in Europe—because such governments would seek close relations with Russia. A paranoid liberal conspiracy theorist (which I am not) would therefore have to assume that the Green Party program was drafted with the help not only of Marine Le Pen but also of Vladimir Putin.

I have chosen to pick on the French Greens because of the particular floridity of their language; but all the Green parties in Europe share the same positions on mass migration and nuclear energy.[7] As Chapter 2 argued, mass migration is indirectly bad for climate change action because by dividing Western societies it makes the creation of stable coalitions against climate change more difficult, and because it strengthens the vote for parties that deny climate change and oppose action. But calls for the abolition of nuclear power harm action against climate change in the most direct possible way; and in Germany and elsewhere, they have had a very serious and damaging effect on policy.

As James Lovelock, David Wallace-Wells, and others have pointed out, if climate change is indeed the threat to modern civilization that the environmentalists (including themselves) believe, then they should be prepared to accept nuclear energy as one of the quickest and most effective alternatives to fossil fuels as far as electricity generation is concerned, and a much more limited threat to human health and well-being. They most certainly should not be seeking to abolish the nuclear power plants that already exist.[8]

The only direct deaths from a nuclear accident were when the reactor at Chernobyl in the Soviet Ukraine exploded in 1986, and numbered 30, with 14 more deaths among Chernobyl workers later. The United Nations Scientific Committee on the Effects of Nuclear Radiation (UNSCEAR) estimates that there have been fewer than 100 deaths since then that can definitely be attributed to the Chernobyl accident, though up to 2005 there were around 6,000 cases of thyroid cancer among people who were children or adolescents in the region in 1986.[9] A series of scientific surveys have suggested that the increase in local cancer deaths as a result of the accident at the Three Mile Island nuclear plant in Pennsylvania in 1979 has been statistically insignificant.[10] As David Wallace-Wells has written, compared not just to the casualties threatened by climate change, but those who already die from pollution due to fossil fuels, "the scars [of nuclear energy] are almost phantom ones."[11] At least nine *million* human beings die every year as a direct result of air pollution caused by fossil fuels. From accidents alone in the global coal and oil industries, around 45,000 people died between 1969 and 2000.[12]

Nobody died as an immediate result of the Fukushima accident that took place as a result of the Tohoku earthquake of 2011 (except for suicides as a result of psychological stress, but this was largely caused not by the accident itself, but the earthquake and tsunami, and the effects of the evacuation of the local population). The World Health Organization estimates that there may be a one percent higher long-term chance of developing cancer among the local population as a result of the Fukushima accident.[13] Yet this accident drove the German government to phase out nuclear power—in a country which does not lie in an earthquake zone and in all recorded history has never suffered from an earthquake or tsunami! The result is that Germany, which was previously doing relatively well in reducing greenhouse gas emissions because of strong state support for renewables, is now set to miss its targets under the Paris agreement. When pressed on the long-term effects on Germany even of the Chernobyl disaster, the best that former German Green Party chair Claudia Roth could come up with was that

30 years later, there are still restrictions in parts of Germany on gathering wild mushrooms and eating wild boar.[14] And that is only due to extreme German official caution in these matters. By some estimates, you would in fact have to eat around 13 kilos of wild boar at a sitting in order to receive the same dosage of radioactivity as from a transatlantic flight.[15] When set against Germany's responsibility for climate change, such an answer is morally frivolous. Especially in Germany, the Greens' hostility to nuclear energy has severely delayed the reduction of reliance on fossil fuels for which they have been struggling. They too have not understood the meaning of the world "priority," or that in this case the fight against climate change requires the assumption not just of limited economic sacrifices but also of limited physical risks.[16]

The platform of the German Greens of course includes climate change, but only as one among a host of other issues, including the abolition of nuclear power. On international cooperation, the Greens emphasize their opposition to "Realism" but adopt ideological positions on the promotion of human rights and democracy that make real cooperation with China and Russia extremely difficult. The stress on "common values" with other nations is really simply a demand that they accept the Green version of Western values.

On migration and citizenship for migrants, the Greens have adopted a position that is digging a deeper and deeper gulf between left and right in Germany, at a time when the threat of climate change should be uniting them. All over Western Europe the Greens and new radical left-wing parties are increasingly drawing support away from the Social Democrats. Meanwhile, populist nationalist parties like Alternative for Germany (Alternative für Deutschland [AfD]) are doing the same to the Christian Democrats.

The gulf in German and European politics therefore risks widening to a point where politics and government become paralyzed in the way that has happened in the United States—with very dangerous results for Europe in general and the struggle against climate change in particular.[17] The echo of Weimar is a real one—not because anything as terrible as the Nazis is waiting in the wings (in this generation at least) but because as in the last years of Weimar, we may be heading into a situation where the creation of stable parliamentary majorities becomes mathematically impossible.

The Christian Democrats in Germany have not done nearly enough to live up to their promises to limit CO_2 emissions, but at least they recognize anthropogenic climate change and their duty to do something about it. Like populist nationalists across Europe, AfD has adopted the climate change denial line of the US Republicans—an insane position for AfD to take, given their terror of international disorder and mass migration.[18] The growing vote for AfD has not been driven by denial of anthropogenic climate change but overwhelmingly by anxiety about mass migration and the cultural transformation of Germany—but the results for German action on climate change if AfD continues to grow will be disastrous.[19]

Chapter 1 described and critiqued what I called "residual elites" in Western militaries and security establishments. What the example of the European Greens illustrates is that there are also residual counter-elites. The attitudes of the Greens to migration were formed on the left when mass migration from outside Europe did not yet exist. Their attitudes to nuclear energy derive from the campaigns during the Cold War against nuclear *weapons*—campaigns that were noble in their intentions but failed totally (and inevitably) in their goals. Somehow the Greens seem to think that the campaign against nuclear energy is a continuation of and a compensation for these failed campaigns of the Cold War past. It is rather as though, having failed in a campaign to ban the use of aircraft in war, the Greens have moved on to a campaign to ban the use of aircraft in everything.

This residual factor also applies to wider attitudes among radical Greens. Thus I am fully in agreement with the title of Naomi Klein's book on climate change, *This Changes Everything*.[20] The problem is that among the things that it has not in fact changed is Klein's own ideological priorities, which remain almost exactly what they would have been if climate change did not exist.[21]

Climate change deniers on the right, like most contemporary Republicans in the United States, are therefore partly correct when they allege that a good deal of agitation against climate change on the left is in fact driven by hostility to capitalism. This also helps explain the blind opposition of most Greens to even researching the possibility of climate change mitigation and carbon removal—because this would remove one argument for the elimination of capitalism.

But of course the Republicans are also completely wrong. The World Bank, the World Economic Forum, Lord Stern, Jeffrey Sachs, Tony Blair, Angela Merkel, Emmanuel Macron, Barack Obama, and John Kerry are not communists. Nor (if he will forgive my saying so) is President Xi Jinping. They have been convinced by the scientific evidence and the scientific consensus. And when it comes to saving capitalism by reforming it, Otto von Bismarck, Theodore Roosevelt, Franklin Delano Roosevelt, Conrad Adenauer, and Charles de Gaulle were also not communists.

The Green New Deal

In the United States, the Green New Deal resolution of January 2019 presented in the House of Representatives by Alexandria Ocasio-Cortez and Ed Markey managed to avoid most of the follies of the European Greens while falling headlong into other holes of the radical Democrats' own digging.[22] The resolution did, however, have the tremendously beneficial effect of forcing parts of the Democratic Party establishment to start taking climate change and action against it really seriously. As a result, the Green New Deal has moved to the center of the Democratic platform.

And Lord knows the Democratic establishment needed galvanizing. Until 2019 their record was infinitely better than that of the Republicans—though that isn't saying very much. But given the importance and urgency of the issue, the Democrats previously failed badly in their response. Nor can this be put down entirely to the fact that during most of the Clinton and Obama administrations, the Republicans controlled Congress and made it their business to block every measure put forward.

At the start of the Clinton administration in 1993–94, the Democrats controlled Congress, and it was in fact Democratic representatives who forced Clinton to abandon his initial plans for a carbon tax in favor of limited, voluntary, and "market-based" strategies—which failed to reduce emissions to anything resembling the required degree. The Byrd-Hagel resolution in the Senate blocking the United States from ratifying the Kyoto Agreement was a bipartisan one. The sums of money invested by the Clinton and Obama administrations in research and development of alternative energy appeared large in themselves—but were in fact small compared to the sums invested by China during the same period. Of all leading figures in the Democratic Party at the national level, as of 2018 only Elizabeth Warren had a genuinely strong record on combating climate change. Others have jumped on the bandwagon of the Green New Deal—with what sincerity remains to be seen.

Throughout the past generation, most of the Democratic establishment has been at one with the Republicans in insisting that developing countries sign up fully to every international agreement if the United States is to do so (as demanded by Byrd-Hagel). This has mired the negotiations in endless battles over historical environmental justice; if the goal was really to reduce emissions, it would have been far better to ignore the developing countries (which, apart from China and India, produce a tiny share of emissions and do not seriously threaten US industry) and concentrate on agreement among the handful of countries that are responsible for more than two thirds of greenhouse gas emissions.

As of 2018, there was an impression that the Democratic Party establishment was actually drifting away from climate change action:

> The Democratic Party has yet to craft a unified message of climate action. Instead, the party as a national establishment seems to be moving backward, toward a policy that embraces a staid status quo rather than the kind of forward-thinking necessary to stop the climate crisis. . . .
>
> Thus far, the Democratic Party has mostly failed to embrace climate change as a social justice or economic equity issue. Instead of tying the transition to a clean energy economy to other popular proposals like a federal jobs guarantee or a nationwide investment in infrastructure and attendant job creation, the party has clung to the status quo of an all-of-the-above energy platform and a conception of climate action that seems to begin and end with market-based solutions like a carbon tax.[23]

This impression was considerably strengthened by the decision of the Democratic National Committee in August 2018 (by a vote of 30 to 2) to reverse its previous stance and go back to accepting donations from "fossil fuel workers and their unions *or employers' political action committees*" (my italics). Announced as a move to restore ties to the trade unions (in an area that is only 4 percent unionized), this was in fact an absolutely transparent attempt to raise money from the energy industry and its lobbies—with consequences for genuine commitment to tackle climate change that hardly need spelling out.[24]

As to the Green New Deal, the idea of using a switch to alternative energy and energy conservation as a way of driving technological revolution and industrial regeneration has been around for a generation. The phrase itself began to be widely used from 2006 on. Arnold Schwarzenegger talked a great deal about a Green industrial revolution as governor of California from 2003 to 2011.[25] The Green New Deal Group in Britain was founded in 2007.[26] As working-class discontent and radicalization gathered pace with the recession of 2008, so support grew for the need to add strong new social programs to the mixture.

Yet for more than a decade, only candidates of the Green Party actually stood on this platform in US presidential elections. As of the end of 2018, only two candidates—Elizabeth Warren and Jay Inslee—had adopted the Green New Deal idea and made combating climate change a key part of their platforms. Only these two had incorporated recommendations from the *Briefing Book for a New Administration* on climate change and national security (issued by the Climate and Security Advisory Group, including a number of distinguished retired generals and admirals) in their thinking.[27]

For that matter, on the national strategic priority of incorporating energy conservation in a program of rebuilding US infrastructure, this had been put forward years earlier by two military officers, Captain Wayne Porter USN and Colonel Mark Mykleby USMC (encouraged by Admiral Mike Mullen, then chairman of the Joint Chiefs of Staff)—and rejected by the Obama administration.[28]

Initially, the Democratic establishment tried to brush off the Green New Deal resolution with some version of Nancy Pelosi's "Green Dream or whatever" remark. By the summer of 2019, however, every leading Democratic candidate had adopted the basic idea in their platforms and they were falling over themselves to promise trillions of dollars for the project. For the first time, leaders of the Democratic Party were beginning actually to respond to the challenge of the Intergovernmental Panel on Climate Change (IPCC), that effective action against climate change will require "rapid, far-reaching and unprecedented changes in all aspects of society," and Lord Stern's warning that otherwise, "there is a big probability of a devastating outcome."

All of the candidates' platforms called for concrete measures to cut emissions to zero by 2050—something that implies doubling the rate of reduction promised by the Obama administration. Even some Republicans were being forced to respond.[29] Moreover, for the first time, this idea was receiving really serious discussion in the mainstream US media. So Ocasio-Cortez, Markey, and their supporters have made a very important contribution to the movement against climate change.[30]

The most admirable things about these new plans by leading Democrats are that for the first time they fully link action against climate change to radical reforms of the US economy and social policy, of a kind that are becoming more and more necessary; that they link the fight against climate change to technological renewal of the US economy; that they all focus attention on generating jobs; and that by appealing to the memories of the New Deal and the Second World War, they remind Americans of great collective national efforts in the past, and that once upon a time Americans could successfully carry out sweeping and successful transformations of their economy and society through state action.

All of the leading Democratic platforms talked of the need to rebuild American infrastructure—including the restoration of US railways—and create infrastructural resilience.[31] They also talked of the need to compensate workers in the fossil fuel sector (and especially coal miners) for the loss of their employment—something that was sadly missing from the Green New Deal resolution.[32] Hopefully this will mobilize organized labor behind the Green New Deal.[33]

All of the leading Democratic platforms for the 2020 elections emphasize the threat of climate change to national security and refer to military assessments of the threat to US bases and the fact that the US military has itself long recognized the reality and danger of climate change.[34] All refer to Cold War images of security-led investment in technological innovation. The Democrats, however, need to do more to exploit the growing bipartisan commitment to rivalry with China in a positive way by stressing how far and fast China is pulling ahead of the United States with regard to investment in infrastructure and research and development. Even such devoted supporters of American-style free-market capitalism as Thomas Friedman have come to recognize that this country is failing badly compared to China when it comes to new infrastructure.[35]

Here is a good opportunity for the Green New Deal to exploit American patriotic sentiment and bipartisan national security concerns—and mobilize some retired military figures in public support. The need for "a new Sputnik Moment"—when the United States woke up to the USSR's growing lead in space exploration—has become a frequent trope in environmentalist writing calling for the country to invest far more money in the research and development of alternative energy.[36]

It is very important that all the Green New Deal platforms of the Democratic candidates in 2020 are optimistic and positive in tone. They emphasize the great chances for economic and social *renewal* that the fight against climate change offers to America. This may seem an odd thing to say in a book that has used the words "catastrophe" and "disaster" quite so many times, but it is US electoral politics 101. All the great American political communicators of the 20th century—Theodore Roosevelt, FDR, Dwight D. Eisenhower, Martin Luther King Jr., and Ronald Reagan—understood that to win Americans' hearts it was necessary to present an uplifting vision of a happier future.[37] In the depths of the Great Depression, with unemployment at more than 25 percent, FDR's campaign song was "Happy Days Are Here Again."

In order to create a "new dispensation" in American politics and political culture, what the Democratic Party still needs to do is to frame the vision, the needs, the positive goals, and the public appeal of the Green New Deal not just in socioeconomic terms but in truly *national* and nationalist ones; to infuse it with "a conscious sense of national purpose," and in turn, to build this national sense of purpose and pursuit of the common good around the struggle to limit climate change.[38]

Above all, the Democrats need to heed words like those of a lifelong Democratic voter: "I voted for Reagan in the 80s because he made me happy and proud to be an American."[39] Nothing has hurt the Democrats more among many former Democratic voters than the perception that they are not proud to be Americans—something that is certainly not true of the Democratic leadership but is true of many radical Democratic activists. Every opportunity should be seized for presenting the Green New Deal as a great national project aimed not only at helping working-class Americans but also at wresting from Donald Trump and his followers the promise to make America great again.

New Dispensations

In order to implement the Green New Deal, the Democrats will of course need to win elections—and not just win them, but win them and keep winning them by sweeping majorities (partly in order to dominate the Supreme Court); and includes not just the presidency but also most congressional seats and governorships. In the words of one of the Green New Deal's supporters, "According to the latest science, we need to eliminate virtually all greenhouse gas emissions by 2050 at the latest, and the projects explained below will take 20 years, so a Green New Deal needs to engender the kind of overwhelming support from the public that can withstand years of propaganda from the fossil fuel industries, right-wing billionaires and conservatives in general."[40]

If they are really to be implemented, and not forgotten—like so many of the campaign promises of the Clinton and Obama presidencies, alas—the ecological, social, and economic goals of the Green New Deal will all require reforms on a scale not seen in the West since the 1940s, and they will need to be maintained steadily over several decades. To achieve these, it will not be enough to achieve a slim electoral majority in the US Electoral College or a fractious and fragmented coalition (as in Europe), or even simply to win large national majorities in individual elections. It is necessary to create a new consensus on basic issues extending across most of the political spectrum that will remain in place even when a Republican or other conservative government takes power.

The lessons of US history over the past century are very clear in this regard. Franklin Delano Roosevelt and Ronald Reagan did not just win elections and form administrations. They created "new dispensations" that lasted for a generation and more. From 1932 to 1980, the legacy of the New Deal and its regulated version of capitalism achieved political and ideological hegemony in the United States. It was accepted by the Republican administrations of Eisenhower and Nixon as well as Democratic administrations. Opponents were relegated to the political fringes or soundly defeated in elections. President Eisenhower declared in the 1950s, "Should any political party attempt to abolish social security and unemployment insurance and eliminate labor laws and farm programs you would not hear of that party again in our political history. There is a tiny splinter group of course that believes you can do these things. Among them are a few other Texas oil millionaires and an occasional politician or businessman from other areas. Their number is negligible and they are stupid."[41]

From 1980 to—let us hope!—2020, the boot was on the other foot. Ronald Reagan achieved a hegemony of unregulated capitalism that has dominated US politics for the past four decades. Just as Eisenhower, Nixon, Macmillan, and Heath accepted the consensus founded by FDR and Attlee, so Clinton, Obama, and Blair accepted the one created by Reagan and Thatcher. Indeed, the Democratic leadership collaborated willingly in the New Deal's destruction.[42]

The foundations of these hegemonies were laid by massive electoral shifts. Franklin Delano Roosevelt won the elections of 1932 and 1936 by 57 percent to 39 percent (a Democratic wave that also resulted in 72 percent of the House of Representatives and 61 percent of the Senate) and 60 percent to 36 percent, respectively. Reagan in 1980 won by 50 percent to 44 percent, but to Reagan's vote can also be added 6 percent for an independent Republican candidate. In 1984 he won by 59 percent to 41 percent. Those are the kinds of figures the Democrats need to aim at if they are going to achieve the hegemony of a Green New Deal over the next four decades. These figures were achieved not through party loyalism but by appealing to voters who had previously been on the other side. FDR won by attracting millions of former Republican voters. Reagan famously won by attracting former New Deal Democrats.

The challenge faced by supporters of the Green New Deal in this regard is significantly greater than that faced by FDR. In the first place, the 1932 elections took place at a time when the economy as a whole was in deep collapse, and even many capitalists realized that American capitalism had gone seriously wrong. Today, the situation is more comparable to that faced by the Progressive movement a generation earlier. The economy has been growing very nicely—but the benefits of that growth have been grotesquely distributed; and the social, cultural, political, and environmental side effects have been dire.

The United States at the time of the original New Deal (with the exception of the tragic and monstrous oppression and exclusion of blacks) was considerably more united than it is today, as a result of the successful integration into American society of the great waves of European immigrants who arrived from the 1870s to 1914. Quite apart from economic constraints, present and future economic circumstances in the West also make the re-creation of general material prosperity extremely difficult. Success in building and sustaining a national majority behind the Green New Deal will depend largely on the creation of new forms of what Jared Diamond has called national "ego strength," and Garrett Hardin called "moral capital": "The degree to which a community possesses interlocking sets of values, virtues, norms, practices, identities, institutions and technologies that mesh well with evolved psychological mechanisms and thereby enable the community to suppress or regulate selfishness and make cooperation possible."[43]

Nonetheless, the goal of a new national dispensation ought to be within the Democrats' grasp. At the elite level, the developments of recent years have seen a widespread shift of intellectual opinion away from the "Washington Consensus" of unlimited free-market capitalism and toward policies of regulated capitalism and social solidarity. This has included former orthodox free-market economists like Larry Summers as well as right-wing populists like Tucker Carlson.

As far as Republican voters are concerned, since 2016, Trump has betrayed almost every promise made to his ordinary voters (with the partial exception of his stance against illegal immigration). Trump's promises to "drain the swamp" and break with Washingtonian cronyism became (predictably) a fantastically bad joke. His promises to rebuild US infrastructure and create millions of well-paid jobs evaporated due both to a complete lack of genuine interest on the part of his administration and the fact that his tax cuts for the rich destroyed any possibility of paying for such a program. If the Democrats cannot win masses of former Democratic voters back again and keep them, then there must be something seriously wrong with their appeal.

Green Intersectionality

Unfortunately, there *is* something seriously wrong with some of their appeal as far as such voters are concerned, and this was reflected in the Green New Deal resolution of Ocasio-Cortez and Markey. So it is just as well (ungrateful though it may seem to say so) that much of its language was omitted from the platforms of the leading Democratic candidates for president. Above all, the Democrats cannot afford to be tainted by the atmosphere of blanket hatred of core American traditions that suffuses their most radical supporters.

The Ocasio-Cortez–Markey resolution was essential to galvanize the mentally and morally exhausted Democratic Party oligarchy into action; but it also contains elements that would make the achievement of a new national dispensation based on the Green New Deal quite impossible. This is because the resolution is framed in the language of "Green Intersectionality," as reflected in the following passage, which is repeated almost word for word twice:

> Whereas climate change, pollution, and environmental destruction have exacerbated systemic racial, regional, social, environmental, and economic injustices (referred to in this preamble as "systemic injustices") by disproportionately affecting indigenous peoples, communities of color, migrant communities, deindustrialized communities, depopulated rural communities, the poor, low-income workers, women, the elderly, the unhoused, people with disabilities, and youth (referred to in this preamble as "frontline and vulnerable communities") . . .

These passages in the Green New Deal resolution derive from "Green Intersectionality." This is an ecological extension of intersectionality in general, a highly influential theory on the left, which holds that alleged discrimination against women (including wealthy and powerful women) and gays (including wealthy and influential gays) is closely akin to racial oppression, and that all these different identities can be mobilized into one movement for their common "liberation." The unintended effects of this are greatly to downplay social and economic (as opposed to sexual or racial) disadvantage, and to class the most economically and socially disadvantaged white males among the privileged oppressors.[44] The political effects have been predictably disastrous. Try telling the unemployed men of a ruined midwestern steel town that they are more privileged than Hillary Clinton or Oprah Winfrey and then ask them to vote for a Democrat.[45]

Why, for example, did the resolution have to slip in a completely gratuitous (and in spirit, mendacious) insult to the white working classes by claiming "a difference of 20 times more wealth between the average white family and the average black family"? The *average* is of course grotesquely skewed by the enormous share of wealth (42 percent of the total) held by the (overwhelmingly white and Asian) top 1 percent. The average says nothing at all about the (stagnant or declining) incomes of the bottom 50 percent of the white population.[46] Why not stress instead that 90 percent of the US population—of all races—have derived no additional benefit from the US economic growth of the past 20 years? Or that the bottom third—of all races—have seen their incomes decline since the 1980s?

This sort of language used in the Green New Deal resolution is politically disastrous because it gives yet more opportunities to the Republicans to tell white working-class voters that the Democrats are not interested in them or are actively hostile to them, and that any benefits from the Deal will go to minorities. It is also highly doubtful that it will win the Democrats a single extra minority vote.

Hard-line cultural liberal positions are not even popular with most Democrats. According to a poll of 2018 by More in Common, almost 80 percent of both blacks and whites in the United States dislike political correctness, even while most (75 percent) also say that hate speech is a problem and have come to approve gay marriage (58 percent) and other liberal causes. "Progressive activists" made up only 8 percent of those polled, and "devoted conservatives" only 6 percent. This poll described 67 percent of Americans belonging to what it called an "exhausted majority," not fully subscribing to either hard liberal or conservative positions and opposed to further polarization.[47]

In a striking finding, only 30 percent of blacks polled consider themselves liberal (reflecting in part strong black religious traditions). If many blacks stayed at home during the 2016 elections instead of voting for Hillary Clinton, it was not because her rhetoric was not "intersectional" enough. It was because they had not forgotten the way in which her husband's administration in the 1990s slashed social welfare and filled the jails with black prisoners.[48]

As Robert Reich predicted in an eerily prescient book of 2015, in the elections the next year, Wall Street backed Hillary Clinton—and the infuriated white working classes swung to an insurgent against both party establishments.[49] Unfortunately, that insurgent was Donald Trump. The Democrats must not let that combination happen again.

Of course, Democrats have a civic duty genuinely to *help* minorities who will vote for them anyway or not vote at all, but they need to pitch their electoral appeal to public voters who will *not* vote for them without considerable effort on the Democrats' part (and then keep their electoral promises to everybody—unlike the last two Democratic administrations).

From this point of view, it was also deeply foolish to give the impression that Democrats have more respect and affection for illegal immigrants (who can't vote) than American workers (who can). At the core of Democratic and democratic philosophy should be the principle that "the only society that can treat all of its members with respect is one in which every individual enjoys rights on the basis of being a citizen, not on the basis of belonging to a particular group."[50]

The Green New Deal sections of the Democratic platforms for 2020 mention the particular need to help black communities, but they are overwhelmingly pitched at ordinary Americans and especially poor Americans in general. And this make sense not just electorally but morally and practically—if the intention is actually to help poor black communities and not just make loud self-righteous noises. Color-blind programs to reduce poverty will disproportionately help blacks because blacks are disproportionately poor. And the same thing applies to fighting climate change if indeed poor blacks are disproportionately threatened by its effects (as they were in New Orleans during Hurricane Katrina, along the Mississippi valley during the great flood of 1927, and in the aftermath of the Okeechobee hurricane of 1928).

Climate change activism has become associated with the cultural liberals' sacralization of different ethnic and cultural identities and gratuitous attacks on conservative cultural symbols in recent decades, and this has necessarily alienated conservatives who might otherwise have recognized the threat of climate change to their nations, to democracy, and to the capitalist order. In part as a result of this public perception of the association of climate change activism with cultural liberalism, all too many conservative American communities have come to see climate change denial as part of their communal culture; like owning guns and attending church. In these sections of the US population, denial of climate change hardly pretends any more to be evidence-based. It has become a matter of collective identity and loyalty. This cultural dogmatism of the American right on climate change has had a disastrous influence on Canadian and Australian conservatism and is beginning to do the same in Germany, the UK, and other parts of Europe. It even influences Russians, much though they would wish to deny it.

Wooing the Elephant

In a nicely appropriate piece of imagery (since the Republican Party symbol is an elephant), social psychologist Jonathan Haidt writes that "the human mind is divided, like a rider on an elephant—and the rider's job is to serve the elephant," or in other words, "intuitions come first, strategic reasoning second."[51] By this, Haidt means that for the great majority of human beings, judgments and views that they themselves take to be derived from rational thinking based on an examination of evidence are in fact largely driven by preexisting beliefs and worldviews that have a deep reciprocal relationship with fundamental human emotions; and that is as true of most "educated" people as of uneducated ones.

I came to this conclusion myself very unwillingly, and I came to it the hard way: through seven years in Washington, DC, dealing with both Republican and Democratic policymakers and public intellectuals. As a dear friend and mentor, Ambassador Bill Maynes, once told me, "Anatol, you have every qualification to work in the policy world of Washington except one. You still believe in the Enlightenment."

This helps explain why attempts to convince conservative voters of the reality and dangers of climate change through evidence and rational argument have so far had so little success: on this issue, the rational intellect is simply not in charge of their brains (on other issues like immigration it is not really in charge of liberal brains either). It is imperative therefore that environmentalists find ways of appealing to Republican voters' inner elephants.

This emphatically does *not* mean seeking compromises with the Republican Party leadership. As the repeated attempts of President Clinton and President Obama showed, that is a completely pointless strategy (there used to be a rather hopeful tendency called the "crunchy conservatives," but it looks as if Trump ate them). It may be possible to win over a few decent dissidents. But above all, it obviously means trying to win over, or win back, a considerable number of moderate conservative voters.

This means finding the right buttons to push, and the right people, or messengers, to push them. Hence the importance—quite apart from the truth of climate change's dire threat to national security—to enlist military figures to argue for the need to take radical action to limit climate change. As I argue throughout this book, the appeal to national security is perhaps the only way to get through or around the dogmatic free-market capitalist ideology of contemporary Western elites.

Of course, the inner elephant itself is often a profoundly confused animal. When it comes to "conservatives" in the West, two fundamental impulses, or perhaps what Ursula Le Guin called "psychomyths," are in conflict.[52] The first is not a conservative myth at all but a modern progressive one (whether in capitalist or socialist guise): that of limitless human technological, material, and moral progress and growth: the Elon Musk myth, if you will, or *Elephas Maximus Economicus*. Where this instinct is overwhelmingly dominant, there is no point in appealing to the elephant. Here, the effort has to be to convince the rider through reasoned argument and evidence that climate change is a real and terrible threat to Western economies, but one that can be tackled precisely through technological development and guided economic development.

Lurking in the truly conservative corners of the inner elephant's soul—especially some of the religious ones—there is another instinct, one that links it very closely indeed to environmentalist thinking, and which indeed is probably the basic instinctual reason why many people became climate change activists in the first place. This is the tendency, reflected in the most ancient psychomyths like the stories of the universal flood in different cultures, that overweening human arrogance, ambition, greed, and self-indulgence will inevitably be punished by Heaven or the sacred order of nature:

> Were they ashamed when they had committed abominations? Nay, they were not at all ashamed, neither could they blush. . . . I will surely consume them, saith the LORD: there shall be no grapes on the vine, nor figs on the fig tree, and the leaf shall fade; and the things that I have given them shall pass away from them.[53]

Perhaps one of the reasons for the incoherence of much of liberal and socialist environmentalist thought is that at a subconscious or elephantine level of their brains, these two psychomyths—on the one hand, limitless progress and improvement; on the other, strict limits set by nature—run into each other. One thing is certain. If climate change activists wish to create solid, dominant, and enduring political coalitions behind action against climate change, they cannot afford deliberately and gratuitously to push the kind of elephantine buttons that will make this impossible.

The growing immiseration of large sections of the white working classes is opening up important new political possibilities across racial lines—if the Democrats know how to use them. For among those conservatives who really are conservatives as opposed to extreme free-market liberals, and who actually care about their white working-class base (unlike both Donald Trump and his "respectable" establishment Republican opponents), something very significant is beginning to happen.

For many years, the social problems of poor black communities were attributed by both hostile and sympathetic commentators to specifically black social and cultural features: by white racists to black inferiority, by others to white oppression and legacies of slavery. What some commentators on the right have begun to see is that exactly the same patterns of family breakdown, single motherhood, drug addiction, domestic abuse, and crime are growing in de-industrialized and newly impoverished white communities.[54] In other words, the causes are common ones that operate across racial lines: poverty, unemployment or highly insecure and temporary jobs, isolation, and social despair. When it comes to the destructive effects of unconstrained free-market capitalism on working-class families, right-wing commentator Tucker Carlson is beginning to sound rather like Michael Moore:

> Thirty years ago, conservatives looked at Detroit or Newark and many other places and were horrified by what they saw. Conventional families had all but disappeared in poor neighborhoods. The majority of children were born out of wedlock. Single mothers were the rule. Crime and drugs and disorder became universal. . . . Virtually the same thing has happened decades later to an entirely different population. In many ways, rural America now looks a lot like Detroit.
>
> Republican leaders will have to acknowledge that market capitalism is not a religion. Market capitalism is a tool, like a staple gun or a toaster. You'd have to be a fool to worship it. Our system was created by human beings for the benefit of human beings. We do not exist to serve markets. Just the opposite. Any economic system that weakens and destroys families is not worth having. A system like that is the enemy of a healthy society.[55]

This statement is especially striking for me because when I was at the Carnegie Endowment for International Peace in Washington in the early 2000s, I was told by senior colleagues from the liberal establishment that I should never use the words "capitalism" or "capitalist" in public because "they sound hostile" and "Americans don't like to hear them." I must always use "free market."[56] Conservative analysts like Yuval Levin now talk of the "alienation" of the white working classes—in origin, a Marxist term.[57]

Ultra-free-market commentators on the right for their part are beginning to direct at poor white communities the same ruthless cultural contempt that they have always used against poor blacks.

> If you spend time in hardscrabble, white upstate New York, or eastern Kentucky, or my own native West Texas, and you take an honest look at the welfare dependency, the drug and alcohol addiction, the family anarchy—which is to say, the whelping of human children with all the respect and wisdom of a stray dog—you will come to an awful realization. . . . The truth about these dysfunctional, downscale communities is that they deserve to die. . . . The

> white American underclass is in thrall to a vicious, selfish culture whose main products are misery and used heroin needles.[58]

And this passage, though especially harsh, only echoes the steady rain of open contempt that the white upper classes and liberal intelligentsia have poured on the white lower classes over the past generation. As Barbara Ehrenreich has warned,

> It's easy for the liberal intelligentsia to feel righteous in their disgust for lower-class white racism, but the college-educated elite that produces the intelligentsia is in trouble too, with diminishing prospects and an ever slipperier slope for the young. Whole professions have fallen on hard times, from college teaching to journalism and the law. One of the worst mistakes this relative elite could make is to try to pump up its own pride by hating those—of any color or ethnicity—who are falling even faster.[59]

The immiseration of the white working classes produced the phenomenon that (according to polls) between 6 and 12 percent of Trump voters previously supported Bernie Sanders, and a quarter of Trump supporters also sympathize with Sanders.[60] These are votes that supporters of the Green New Deal can and must win.

If there is one issue that cuts across lines of race and class in the United States and elsewhere, it is climate change. If there are three issues that should unite poor blacks and whites, they are the regeneration of America's industrial base, the provision of an adequate safety net for all Americans in terms of health care and social security, and support for families with children. As that great American social democrat and civic nationalist Martin Luther King Jr. argued in his last book, shortly before he was assassinated,

> A substantial group [of White Americans] is composed of those having common needs with the Negro, and who will benefit equally with him in the achievement of social progress. There are, in fact, more poor White Americans than there are Negro. Their need for a war on poverty is no less desperate than the Negro's. . . . [T]here is no separate black path to power and fulfillment that does not intersect

> white paths, and there is no separate white path to power and fulfillment, short of social disaster, that does not share that power with black aspirations for freedom and human dignity.[61]

Or radical Democratic activists can go on jabbing and jabbing at working-class white voters with talk of "white privilege" and "Anglo supremacy" and "heteronormative families" when if they were to talk of "wealthy privilege" or "elite privilege" or "corporate privilege" or "working families" they could win their votes.[62] I know of no evidence to suggest that intersectionality, political correctness, the Woke movement, and so on have won a single vote for the Democrats that they would not have won anyway. We can't say for sure how many they have lost to the Republicans, but we know that extreme right-wing commentators like Bill O'Reilly, Rush Limbaugh, Sean Hannity, and Mike Savage think it's a lot—because they quote them endlessly on their repellent talk shows. Is the main political achievement of the American left going to be to generate votes for the American right?

When it comes to the Green New Deal, its advocates need to follow and expand on Elizabeth Warren's approach in framing it as a great national and patriotic struggle to strengthen the United States of America.[63] They could frame their platforms—completely sincerely by the way—with speeches like the following on the environment: "To waste, to destroy, our natural resources, to skin and exhaust the land instead of using it so as to increase its usefulness, will result in undermining in the days of our children the very prosperity which we ought by right to hand down to them."[64] And the following on social justice: "Every gun that is made, every warship launched, every rocket fired signifies in the final sense, a theft from those who hunger and are not fed, those who are cold and are not clothed. This world in arms is not spending money alone. It is spending the sweat of its laborers, the genius of its scientists, the hopes of its children. This is not a way of life at all in any true sense. Under the clouds of war, it is humanity hanging on a cross of iron."[65]

Al Gore and Karl Marx? No, President Theodore Roosevelt (Republican) and President Dwight D. Eisenhower (Republican); and the platform could feature giant photographs of them: Roosevelt in his Rough Rider uniform, and Eisenhower in the uniform he wore when he led the Allied armies to victory in Western Europe in 1944–45.[66] Such an approach would at least wrong-foot the Republicans.

Looming behind the question of whether Democrats can win elections and begin to implement a Green New Deal are the probable combined shocks of the next decades, summed up in Chapter 2. It cannot be emphasized too strongly: Western democracies that continue their existing levels of social inequality, national division, and political polarization will not survive these shocks. The challenge set by Herbert Croly more than a century ago is once again true today:

> Just in so far as Americans timidly or superstitiously refuse to accept their national opportunity and responsibility, they will not deserve the names either of freemen or of loyal democrats. There comes a time in the history of every nation, when its independence of spirit vanishes, unless it emancipates itself in some measure from its traditional illusions; and that time is fast approaching for the American people.[67]

The Chinese Communist Party argues that Western democracies are incapable of taking effective action against climate change and that only wise and firm authoritarian governments can do so. Certain Western authors like David Shearman and Joseph Wayne Smith agree.[68] The Green New Deal is a chance to prove them wrong. Not the very last chance—but we are getting awfully late.

Conclusions

I call heaven and earth to record this day against you, that I have set before you life and death, blessing and cursing: therefore choose life, that both thou and thy seed may live.

—Deuteronomy 20:19

IN THE CONTEXT OF the growing impact of climate change and other problems, Western civic nationalism should shift from the expansion to the defense of Western liberal democracy, and from an ethic of conviction to an ethic of responsibility among Western policy elites.[1]

Since the end of the Cold War, the philosophy of Western foreign policy has been dominated by liberal internationalism: an activist and expansionist attitude to spreading Western democracy and Western values across the globe. This has been portrayed as anti-nationalist. In fact, however, it reflects what I have called Western "civilizational nationalism," with certain analogies to the identities of the Roman and Chinese empires; and like them, a strong implication that anyone who does not share our values is ipso facto a barbarian.[2]

This program has failed, as the briefest look around the contemporary world should demonstrate.[3]

It has failed because on key issues, climate change first among them, the United States and the European Union failed to live up to their own self-assigned global missions and provide enlightened, disinterested, and courageous global leadership.

A particularly egregious example has been Western policy toward Libya. Having embraced the Ghaddafi regime for profit, Britain, France, and the United States then overthrew that regime with no plan for how to replace it (in the process, undermining international legitimacy by cynically manipulating a UN resolution to gain spurious legal cover). After loudly proclaiming a Western humanitarian duty to intervene, the West has met the resulting civil war with a combination of indifference and division, with different countries supporting different Libyan factions. Given such examples, why on earth should anyone else in the world trust Western good faith, Western professions of morality, or even Western common sense?

The Western globalization program has failed because like all hubristic projects, it carried within it the seeds of nemesis (remember "The End of History"?). The plan to build a Western internationalist tower of Babylon has collapsed, and we are left once more with a multiplicity of national voices.[4]

It has failed because of irreducible international and local realities that make the successful adoption of Western liberal democracy impossible.

It has failed because of inevitable and insuperable resistance by regional great powers.

It has failed because it was disastrously compromised by Western geopolitical ambitions and national hatreds, which demanded that Russian, Chinese, Iranian, and other liberal reformers betray the interests of their countries by supporting US foreign policy.

It has failed because it was in practice abandoned whenever US geopolitical interests required this, fatally undermining its credibility among the populations it was meant to win over.

It has failed because it was associated with a deeply flawed project of unrestrained capitalist globalization (the "Washington Consensus"), which only worked when it was heavily qualified, controlled, and reshaped by strong, able, and progressive local states. As a result, China has emerged as an alternative model of capitalist development.

It has failed because Western states refused to imitate the East Asians and reshape globalization in the interests of their own populations, through economic planning, controls on financial flows, and restrictions on trade and migration. As a result, Western democracies now suffer from such deep internal problems that their appeal to people in other countries has been radically diminished.

A central reason for the last failure is summed up in a conversation reported by David Goodhart in two of his books. In 2011, Sir Gus O'Donnell, then the most senior British civil servant, told him, "When I was at the Treasury I argued for the most open door policy to immigration. . . . I think it's my job to maximise global welfare not national welfare." A British citizen might well ask: did this man not swear an oath of loyalty to the British people, as represented in the person of the British monarch? Was he not paid out of the taxes of the British people? Did he not receive a knighthood and other honors from the British crown?

A wider issue of responsibility is also involved, as exemplified in the remark of Sir Gus's neighbor at the dinner, Mark Thompson, then director-general of the BBC (his salary also paid out of British taxes), who said that "he believed global welfare was paramount and that therefore he had a greater obligation to someone in Burundi than to someone in Birmingham."[5]

The British elites know very little indeed about Burundi or indeed the great majority of other countries around the globe, and only a bit more about Birmingham in many cases. Even in the cases of countries of great importance to Britain, like Russia and Pakistan, their knowledge is often not just inadequate but wrong. Even in the case of the government department tasked with aiding foreign countries (the Department for International Development, or DIFID), its officials often lack adequate, detailed, and relevant knowledge of the countries they are dealing with, in part because their career paths require them to hop from country to country.

They are also not responsible for these foreign countries' policies or these countries' fates. Why should they be? After all, they do not control or serve in the countries' governments. As a result *they cannot be held accountable*; and that is of course equally true of civil servants and ministers back in London and Washington. With the partial exception of the Balkans, over the past generation the cases where Western powers have taken over countries by force and sought to reshape them have been uniformly disastrous; and it is notorious that nobody, but nobody, has been held accountable.[6]

For the officials (elected or unelected) of a democratic state to hold office without responsibility or accountability violates fundamental principles of democracy. It is in fact what we like to say about officials in Russia, Pakistan, or wherever. It is not just that it risks lapsing into the empty, self-flattering, feel-good pseudo-humanitarianism that has been a moral blight on the contemporary West; as Goodhart suggests, it can also lead to profound indifference to problems at home—for which officials can, and should, be held accountable by the citizens who pay their salaries. The result is elites who are not really responsible to anyone and who therefore lack both the ability and the moral right to ask for commitments and sacrifices from ordinary citizens. As Stephen Holmes and Cass Sunstein have pointed out:

> Under American law, rights are powers granted by the political community. . . . When they are not backed by legal force, by contrast, moral rights are toothless by definition. Unenforced moral rights are aspirations binding on conscience, not powers binding on officials. They impose moral duties on all mankind, not legal obligations on the inhabitants of a territorially bounded nation state. Because legally unrecognised moral rights are untainted by power, they can be advocated freely without much worry about malicious misuse, perverse incentives, and unintended side effects.[7]

From an ethical realist perspective, responsibility to a particular state and nation qualifies but does not abolish wider human responsibilities. However, "While trying to improve evils in men and circumstances which cannot be improved, one loses time and makes things worse; instead, one ought to accept the evils as raw materials and then seek to counterbalance them."[8]

Ethical realism does not deny or contradict transcendental values and goals. On the contrary: by recognizing the limits and constraints of an imperfect humanity in an imperfect world, it is in a position to pursue them far more effectively.[9] Viewed from the perspective of a higher morality, national loyalty may well be a human imperfection. If so, like so many human imperfections, it is an inevitable one, whose good aspects we should turn to good effect; and as Slavoj Zizek has written, "the defence of one's way of life does not exclude ethical universalism."[10]

A commitment to ethical realism therefore means that the first—though not the only—responsibility of officials of a state and elected representatives of a people is to that state and that people. This reflects practicality, as well as law and ethics. As this book has argued, with the exception of saints (who have always been few in number) people's willingness to make sacrifices for distant foreigners is extremely limited. The US public's refusal to sacrifice even very small numbers of American lives on purely humanitarian missions is a good case in point.

The gulf between declarations of global responsibility and inability to mobilize national action has lain at the heart of the failure adequately to address the danger of climate change. Hence the argument of this book: that it is necessary to reframe the struggle against climate change in nationalist terms: the defense of nation states, their interests, and their future survival.

In Western nations, this also means the defense of Western liberal democratic civilization against climate change and its effects: international chaos and internal entropy, division, and decay. This is the duty of every Western citizen, and duty is in the end the strongest barrier against despair—in our case, the cultural, psychological, and political despair and anomie that are eating away at our societies and democratic systems.

This civic nationalist program is defensive but not chauvinist. It requires strict controls on migration but is absolutely committed to racial equality at home, to measures to increase racial equality, and to strategies to assist and integrate immigrant communities. This is a critical part of building the national solidarity that our democracies will need both to limit climate change and to withstand it in the generations to come. Hence also my support for the Green New Deal and its combination of action against climate change, economic regeneration, and social solidarity. Internationally, it is committed to as much cooperation between states as can realistically be achieved, and to the preservation of the European Union—though in a form closer to de Gaulle's idea of a *Europe des Patries* (Europe of Fatherlands).

In terms of the promotion of democracy in the world, this program means shifting Western focus away from propagating democracy and toward strengthening what President John Quincy Adams (warning against interventions abroad) called "the benignant sympathy of

[America's] example"; back to what was in fact the American approach before rising American power and deepening European conflicts led the United States to adopt an activist and missionary strategy of spreading democracy and freedom.[11] It was the obviously superior example of Western democracy and capitalism, not Reagan's "Star Wars," that brought down Soviet communism.

This means that the United States, and the West in general, should concentrate on strengthening our democratic, social, and economic systems at home—and thereby restoring a model that other countries can follow if they wish, without us shouting at them. This also means demonstrably taking the lead on action against climate change. A belief in absolute Western democratic superiority should be qualified by an awareness that if, as seems highly probable, the historians of the future see climate change as the greatest threat to humanity and the greatest test of human political systems, then on performance in this test to date they will grade several leading Western democracies as no better than the Russians, let alone the Chinese.[12] Imagining ourselves as seen through the eyes of future historians is one way of respecting "the requirement of cosmic humility with regard to the moral evaluation of the actions of states."[13]

As this book has argued, concentration on the threat of climate change to all nations both implies and requires the abandonment of moves to a new cold war with Russia and China, and ideological détente and respect for the legitimacy and integrity of each other's political systems.[14] Differences will of course continue, but they should be contained and dealt with on their own terms, not as part of overarching and embedded structures of mutual hostility. Territorial disputes like Crimea and the South China Sea should be treated in the same dispassionate way that we treat other post-imperial territorial disputes in the world. They should be relegated to their proper minor place in the broader scheme of Western national interests, especially compared to the threat of climate change.

When it comes to the threat of climate change, both an approach based purely on short-term national interest *and* an approach based entirely on international humanitarianism would be equally immoral and pointless. Pure humanitarianism is unable to generate the will necessary actually to achieve anything, and results to a great extent in empty hypocritical declarations. An approach based purely on national interest ignores the fact that—as this book has repeatedly stressed—if we do not manage to limit climate change, then in the long run *all* existing nations will be destroyed, some sooner, some later.

This position leads to a resolution of the tension between what have been seen as two diametrically opposed visions of Western responsibility in the face of ecological collapse and over-population: Kenneth Boulding's "Spaceship Earth" and Garrett Hardin's "Lifeboat"—for neither image completely describes the reality they are seeking to conjure up.[15] Boulding is correct that in the long run we are all in the same endangered spaceship; but as Hardin pointed out, spaceships have captains and disciplined crews obedient to orders from above, as well as communities of passengers governed by extremely strong and rigorous communal codes.[16] However, neither of these suppositions is true—or will ever be true—of the human species.

Hardin is correct that to invite any large part of floundering humanity into Western lifeboats would start by causing vicious fights among the existing inhabitants of the boats and end by overturning the boats altogether and drowning everyone. But as far as anthropogenic climate change is concerned a key point is missing. Lifeboats are at the mercy of the wind and waves. Western nations have the power to help calm the storm by limiting climate change and thereby to help save not only themselves but all the others hanging on to life rafts and planks. By reducing our greenhouse gas emissions, we help ourselves and the world. We are not impotent victims of an act of God. Our fate, and that of humanity at large and our descendants in particular, is in our own hands.

Doha and London, 2018–2019

NOTES

Introduction

1. Martin Luther King, *Where Do We Go From Here? Chaos or Community*, introduction by Vincent Harding (reprinted Beacon Press, Boston MA 2010), page 56.
2. For the effect of US-Chinese tensions on climate change negotiations, see Jeff Goodell, "Saving the Paris Agreement," *Rolling Stone*, January 18, 2019, at https://medium.com/rollingstone/saving-the-paris-agreement-4285ac7c83fb—but *Rolling Stone* is not common reading in the US security establishment, alas.
3. Anatol Lieven and John Hulsman, *Ethical Realism: A Vision for America's Role in the World* (Pantheon Books, New York 2006); Hans Morgenthau, *Politics among Nations* (reprinted McGraw-Hill, New York 2005), page 226. See also Anatol Lieven, "Realism and Progress," in *Reinhold Niebuhr and Contemporary Politics: God and Power*, ed. Richard Harries and Stephen Platten (Oxford University Press, New York 2010), pages 169–182.
4. For the concept of the "Anthropocene," see Simon L. Lewis and Mark A. Maslin, *The Human Planet: How We Created the Anthropocene* (Pelican, London 2018).
5. See Lieven and Hulsman, *Ethical Realism*, pages 67–70.
6. Nicholas Stern, *Why Are We Waiting? The Logic, Urgency and Promise of Tackling Climate Change* (MIT Press, Cambridge MA 2015), page 305. See also Mike Hulme, *Why We Disagree about Climate Change: Understanding*

Controversy, Inaction and Opportunity (Cambridge University Press, Cambridge 2009), pages 224–243.

7. Hans J. Morgenthau, *In Defense of the National Interest* (University Press of America, Lanham MD 1982), page 42.
8. See the discussion of this question in David Miller, *On Nationality* (Oxford University Press, Oxford 1995), pages 7–8.
9. Tom Nairn, *Faces of Nationalism: Janus Revisited* (Verso, London 1997); Yael Tamir, *Why Nationalism* (foreword by Dani Rodrik, Princeton University Press, Princeton NJ 2019); Yascha Mounk, *The People vs. Democracy: Why Our Freedom Is in Danger and How to Save It* (Harvard University Press, Cambridge MA 2018); Will Kymlicka, *Politics in the Vernacular: Nationalism, Multiculturalism and Citizenship* (Oxford University Press, New York 2001). For a counter-argument, see Jill Lepore, *This America: The Case for the Nation* (Liveright, New York 2019), page 3 ff.; though curiously enough, having contrasted good "patriotism" with bad "nationalism", Dr. Lepore then goes on to praise good civic "nationalism"! As Lepore writes, "In American history, liberals have failed, again and again, to defeat illiberalism except by making appeals to national aims and ends."
10. Michael Walzer, "Nation and Universe," Brasenose College Oxford, May 1 and 8, 1989, Tanner Lectures on Human Values, page 536, at https://tannerlectures.utah.edu/_documents/a-to-z/w/walzer90.pdf.
11. Paul Collier, *The Future of Capitalism: Facing the New Anxieties* (Allen Lane, London 2018), page 37.
12. Richard B. Howarth, "Intergenerational Justice," in *The Oxford Handbook of Climate Change and Society*, ed. John S. Dryzek, Richard B. Norgaard, and David Schlosbert (Oxford University Press, New York 2011), pages 338–354; Yael Tamir, "Pro Patria Mori! Death and the State," in *The Morality of Nationalism*, ed. Robert McKim and Jeff McMahan (Oxford University Press, New York 1997), pages 227–244. For a readiness for self-sacrifice as a fundamental theme of human culture, see Roger Scruton, *Green Philosophy: How to Think Seriously about the Planet* (Atlantic Books, New York 2013), pages 91–92.
13. Stephen M. Gardiner, *A Perfect Moral Storm: The Ethical Tragedy of Climate Change* (Oxford University Press, New York 2011), pages 143–212, 284–298; Stern, *Why Are We Waiting?*, pages 151–184.
14. Milan Kundera, *The Book of Laughter and Forgetting* (Penguin, London 1980), page 229.
15. Richard Weaver, "The Problem of Tradition," in *A Program for Conservatives* (Regnery, Washington DC 1956).
16. The text of the Green New Deal resolution introduced in the US House of Representatives on February 7, 2019, is to be found at https://www.congress.gov/116/bills/hres109/BILLS-116hres109ih.pdf.

17. Chandran Nair, *The Sustainable State: The Future of Government, Economy and Society* (Berrett-Koehler, Oakland CA 2018), chap. 1.
18. Anthony Leiserowitz et al., "Climate Change in the American Mind: December 2018," Yale Program on Climate Change Communication, at http://climatecommunication.yale.edu/publications/climate-change-in-the-american-mind-december-2018; "Is the Public Prepared to Pay to Help Fix Climate Change," Associated Press–NORC Center for Public Affairs Research, November 2018, at http://www.apnorc.org/projects/Pages/Is-the-Public-Willing-to-Pay-to-Help-Fix-Climate-Change-.aspx; Hal Bernton, "Washington State Voters Reject Carbon Fee Initiative," *Seattle Times*, November 6, 2018, at https://www.seattletimes.com/seattle-news/politics/voters-rejecting-carbon-fee-in-first-day-returns/.
19. Cited in Collier, *Future of Capitalism*, page 59.
20. Roger Harrabin, "Climate Change: Ban Gas Grid for New Homes 'in Six Years,'" BBC, February 21, 2019, at https://www.bbc.com/news/science-environment-47306766.
21. Collier, *Future of Capitalism*, pages 61–62. See also Jeremy Black, *English Nationalism: A Short History* (Hurst, London 2018), pages 188–189.
22. For a liberal internationalist counter-argument, see Simon Dalby, *Security and Environmental Change* (Polity, Cambridge 2009), page 129 ff.
23. David Miller, *On Nationality* (Oxford University Press, New York 1997), page 24.
24. James Cascio, "The Next Big Thing: Resilience," *Foreign Policy*, September 28, 2009, at https://foreignpolicy.com/2009/09/28/the-next-big-thing-resilience/; W. Neil Adger, Katrina Brown, and James Waters, "Resilience," in *The Oxford Handbook of Climate Change and Society*, pages 696–710. See also the site https://www.resilience.org/.
25. See, for example, Karin M. Fierke, *Critical Approaches to International Security* (Polity Press, Cambridge 2015).
26. George Monbiot, *Out of the Wreckage: New Politics for an Age of Crisis* (Verso Books, London 2018), page 49. See also Robert Kuttner, *Can Democracy Survive Global Capitalism?* (W. W. Norton, New York 2018); Karin Backstrand, "The Democratic Legitimacy of Global Governance after Copenhagen," *The Oxford Handbook of Climate Change and Society*, pages 669–684. See also Stephen Holmes and Cass Sunstein, *The Cost of Rights: Why Liberty Depends on Taxes* (W. W. Norton, New York 1999), page 29; Dani Rodrik, *The Globalization Paradox: Democracy and the Future of the World Economy* (W. W. Norton, New York 2011), pages xviii, 19–23, 112–113. For qualified praise for the new medievalism, see Manuel Castells, *Communication Power* (Oxford University Press, New York 2013); Parag Khanna, *Connectography: Mapping the Global Network Revolution* (Weidenfeld and Nicholson, London 2016).

27. David Goodhart, *The Road to Somewhere: The Populist Revolt and the Future of Politics* (Hurst, London 2017).
28. This was clear even before the recession of 2008 led to a new awareness of the need for strong state power to regulate economies. See Samy Cohen, *The Resilience of the State: Democracy and the Challenges of Globalisation,* translated by Jonathan Derrick (Lynne Rienner, Boulder CO 2006); Mounk, *People vs. Democracy*, page 199.
29. Robert Reich, *The Common Good* (Vintage, New York 2019), pages 22–26.
30. Anatol Lieven, *Pakistan: A Hard Country* (Penguin, London and Oxford University Press, New York 2012).
31. David Goodhart, *The British Dream: Successes and Failures of Postwar Immigration* (Atlantic Books, London 2013), page 6.
32. See Sheila Jasanoff, "Cosmopolitan Knowledge: Climate Change and Global Civic Epistemology," in *The Oxford Handbook of Climate Change and Society*, pages 129–143.
33. David Rosenberg, "Ocasio-Cortez Green New Deal Is Surest Way to Lose War on Climate Change," *Haaretz*, May 15, 2019, at https://www.haaretz.com/us-news/.premium-ocasio-cortez-s-green-new-deal-is-surest-way-to-lose-war-on-climate-change-1.6935504.
34. See Erica Buist, "Language Is the Latest Weapon against the Climate Emergency," *One Zero*, May 18, 2019, at https://onezero.medium.com/language-is-the-latest-weapon-against-the-climate-emergency-c64de4b97335.
35. David Wallace-Wells, *The Uninhabitable Earth: A Story of the Future* (Allen Lane, London 2019), page 11.

Chapter 1

1. Raymond Chandler, "Red Wind", in Raymond Chandler, *Stories and Early Novels*, ed. Frank MacShane (Library of America, New York 1995), page 368.
2. In George Orwell, *My Country Right or Left, 1940–43:* Vol. 2 of the *Collected Essays, Journalism and Letters of George Orwell*, edited by Sonia Orwell and Ian Angus (Nonpareil Books, Boston MA 2000), page 75.
3. For an argument in favor of concentrating on the small no. of major emitters, see Gwyn Prins and Steve Rayner, "The Wrong Trousers," a Joint Discussion Paper of the James Martin Institute for Science and Education, University of Oxford, and the MacKinder Centre, London School of Economics, July 27, 2009; Nicholas Stern, *Why Are We Waiting?* (MIT Press, Cambridge MA 2015), pages 23–26.
4. Nathaniel Rich, *Losing Earth: A Recent History* (Farrar, Straus and Giroux, New York 2019), pages 13, 175–294; Kari Marie Norgaard, "Climate Change Denial: Psychology, Culture and Political Economy," in *The Oxford Handbook of Climate Change and Society*, ed. John S. Dryzek, Richard B. Norgaard, and David Schlosbert (Oxford University Press,

New York 2011), page 401. For an overview of climate change denial, see Riley E. Dunlap and Aaron M. McCright, "Organized Climate Change Denial," in *The Oxford Handbook of Climate Change and Society*, pages 144–159.

5. For a good summary of the "debate" on anthropogenic climate change, see Andrew Guzman, *Overheated: The Human Cost of Climate Change* (Oxford University Press, New York 2013), pages 19–53. For a portrait of the denial camp and its roots, see Riley E. Dunlap and Aaron M. McCright, "Organised Climate Change Denial," in *Oxford Handbook of Climate Change and Society*, pages 144–160.
6. *Climate Change 2014: Synthesis Report, Summary for Policymakers*, Intergovernmental Panel on Climate Change (IPCC) (Geneva 2014), page 11, at https://www.ipcc.ch/site/assets/uploads/2018/02/AR5_SYR_FINAL_SPM.pdf. Degree values throughout book are Celsius.
7. See "WMO Statement on the State of the Global Climate in 2018," World Meteorological Organisation no. 1233, 2019, at https://gallery.mailchimp.com/daf3c1527c528609c379f3c08/files/82234023-0318-408a-9905-5f84bbb04eee/Climate_Statement_2018.pdf; Text of Paris Agreement at https://unfccc.int/sites/default/files/english_paris_agreement.pdf; "Is Arctic Permafrost the 'Sleeping Giant' of Climate Change?," NASA Science, June 24, 2013, at https://science.nasa.gov/science-news/science-at-nasa/2013/24jun_permafrost; Chris Mooney, "Arctic Cauldron," *Washington Post,* April 18, 2019, at https://medium.com/thewashingtonpost/what-will-climate-change-do-to-us-3f9968a53dcf.
8. William Nordhaus, "Projections and Uncertainties about Climate Change in an Area of Minimal Climate Policies" (working paper, National Bureau of Economic Research, 2016).
9. See Luke D. Trusel et al., "Nonlinear Rise in Greenland Runoff in Response to Post-Industrial Arctic Warming," *Nature*, December 5, 2018, at https://doi.org/10.1038/s41586-018-0752-4 (2018).
10. See Mark Lynas, *Six Degrees: Our Future on a Hotter Planet* (Harper Perennial, New York 2008), pages 163–241; Daniel B. Botkin et al., "Forecasting the Effects of Climate Change on Biodiversity," *BioScience*, vol. 57, no. 3, March 2007, pages 227–236, at https://doi.org/10.1641/B570306; Carl Zimmer, "The Planet Has Seen Sudden Warming Before: It Wiped Out Almost Everything," *New York Times,* December 7, 2018, at https://www.nytimes.com/2018/12/07/science/climate-change-mass-extinction.html.
11. Will Steffen, "A Truly Complex and Diabolical Policy Problem", in *The Oxford handbook of Climate Change and Society* , ed. John S. Dryzek, Richard B. Norgaard, and David Schlosbert (Oxford University Press, New York 2011), page 27.
12. Jason Treat et al., US Geological Survey, "What the World Would Look Like If All the Ice Melted," *National Geographic,* September 2013; David

Wallace-Wells, *The Uninhabitable Earth: A Story of the Future* (Allen Lane, London 2019), pages 64–68.

13. Wallace-Wells, *The Uninhabitable Earth: A Story of the Future,* page 12.
14. "Risks Associated with Global Warming of 1.5 degrees Celsius or 2 Degrees Celsius," Tyndall Centre for Climate Change Research, University of East Anglia, Norwich, UK, May 2018. For estimates of sea-level rise at different temperatures, see Benjamin P. Horton et al., "Mapping Sea-Level Change in Space, Time and Probability," *Annual Review of Environment and Resources*, vol. 38, 2018, pages 481–521.
15. See the reports of the Mercator Research Institute on Global Commons and Climate Change (MCC) at https://www.mcc-berlin.net/en.html; "Global Warming of 1.5 Degrees," Intergovernmental Panel on Climate Change (IPCC) special report, 2018, at https://www.ipcc.ch/report/sr15/.
16. See Bjorn Lomborg, *Cool It: The Skeptical Environmentalist's Guide to Global Warming* (Knopf, New York 2007).
17. Ian Bremmer, *Us vs. Them: The Failure of Globalism* (Penguin, New York 2018); Paul Collier, *The Future of Capitalism: Facing the New Anxieties* (Allen Lane, London 2018); Dani Rodrik, *The Globalization Paradox: Democracy and the Future of the World Economy* (W. W. Norton, New York 2011), pages 247–248.
18. Robert D. Kaplan, *The Coming Anarchy: Shattering the Dreams of the Post–Cold War Era* (Vintage, New York 2001).
19. See Amitav Ghosh, *The Great Derangement: Climate Change and the Unthinkable* (University of Chicago Press, Chicago 2016), pages 132–134.
20. Roger Scruton, *Green Philosophy* (Atlantic Books, London 2012), pages 14–17, and chapters 5, 10, and 11. See also the US group Conservatives for Responsible Stewardship (CRS) at http://www.conservativestewards.org/about-us/. For much more radical and statist approaches by British Conservative environmentalists, see the Conservative Environment Network (CEN) at https://www.cen.uk.com/.
21. Edmund Burke, "A Letter to a Mmeber of the National Assembly", in *The Writings and Speeches of Edmund Burke, Vol. IV* (Cosimo Classica, New York 2008), page 19.
22. See John A. Mathews, *Greening of Capitalism: How Asia Is Driving the Next Great Transformation* (Stanford University Press, Stanford CA 2015), pages 165, 172, 180ff.
23. For an early statement of this, see the United Nations Development Program (UNDP) Human Development Report of 1994, at http://hdr.undp.org/en/content/human-development-report-1994.
24. Barry Buzan, Ole Waever et al., *Security: A New Framework for Analysis* (Lynne Rienner Publishers, London 1997), pages 26ff., 32–33, 46–47.
25. Colonel Max Brosig et al., "Implications ofClimate Change for the U.S. Army", The United States Army War College, November 2019, at https://

www.americansecurityproject.org/wp-content/uploads/2019/10/Army-Climate-Change.pdf.

26. See, for example, Daniel Deudney, "The Case against Linking Environmental Degradation and National Security," *Millennium: Journal of International Studies*, vol. 19, no. 3, 1990, pages 461–476.
27. Gareth Porter and Janet W. Brown, *Global Environmental Politics* (Westview Press, Boulder, CO, 1991).
28. Marc A. Levy, "Is the Environment a National Security Issue?", *International Security*, vol. 20, no. 2, October 1995, pages 40, 51.
29. For an opposing argument, see Stephen Walt, "The Renaissance of Security Studies," *International Studies Quarterly*, vol. 35, no. 2, pages 211–239.
30. See Olafur Ragnar Grimsson et al., "A New World: The Geopolitics of the Energy Transformation", International Renewable Energy Agency (IRENA), 2019, at http://geopoliticsofrenewables.org/Report; Anatol Lieven, "How climate change will transform the global balance of power", *Financial Times*, November 5, 2019.
31. For an assessment of the impact of climate change on heat waves like that of 2011, see Stefan Rahmstorf and Dim Cornou, "Increase of Extreme Events in a Warming World," Potsdam Institute for Climate Impact Research, March 20, 2012, at http://www.pik-potsdam.de/~stefan/Publications/Nature/rahmstorf_coumou_2011.pdf.
32. Riley E. Dunlap and Aaron M. McCright, "A Widening Gap: Republican and Democratic Views on Climate Change," *Environment*, vol. 50, no. 5, 2008, pages 26–35; Matthew C. Nisbet, "Communicating Climate Change: Why Frames Matter for Public Engagement," *Environment: Science and Policy for Sustainable Development*, vol. 51, no. 2, 2009, pages 12–23; Kari Norgaard, "Climate Denial," *Handbook of Climate Change and Society*, pages 399–413.
33. Susanne C. Moser and Lisa Dilling, "Communicating Climate Change: Closing the Science-Action Gap," in *The Oxford Handbook of Climate Change and Society*, page 165; Matthew Nisbet, "Communicating Climate Change: Why Frames Matter to Public Engagement," *Environment*, vol. 51, no. 2, 2009, pages 12–23; Joseph P. Reser and Graham L. Bradley, "Fear Appeals in Climate Change Communication," in *The Oxford Encyclopedia of Climate Change Communication*, ed. Matthew Nisbet et al. (Oxford University Press, New York 2018), at https://oxfordre.com/climatescience/page/climate-change-communication/the-oxford-encyclopedia-of-climate-change-communication; for the wider theory of framing; see Daniel Kahneman, "Maps of Bounded Rationality: A Perspective on Intuitive Judgement and Choice," Nobel Prize acceptance lecture, December 8, 2002, at https://www.nobelprize.org/uploads/2018/06/kahnemann-lecture.pdf.

34. For climate change and the assessment of risk, see the Centre for the Study of Existential Risk, Cambridge University, at https://www.cser.ac.uk/; David King et al., "Climate Change: A Risk Assessment," Centre for Science and Policy 2015, at http://www.csap.cam.ac.uk/projects/climate-change-risk-assessment/; Paul A. T. Higgins and Jonah V. Steinbuck, "A Conceptual Tool for Climate Change Risk Assessment," *Earth Interactions*, vol. 18, no. 21, 2014. For the cultural and ideological roots of different assessments of risk, see Mike Hulme, *Why We Disagree about Climate Change* (Cambridge University Press, New York 2009), pages 177–211.
35. Nicholas Stern, *Why Are We Waiting? The Logic, Urgency and Promise of Tackling Climate Change* (MIT Press, Cambridge, MA 2015), pages 132–150.
36. Chad Michael Briggs, "Climate Security, Risk Assessment and Military Planning," *International Affairs*, vol. 88, no. 5, 2012.
37. "The Global Risks Report 2019," World Economic Forum (in partnership with Marsh and McLennan Companies and Zurich Insurance Group) at http://www3.weforum.org/docs/WEF_Global_Risks_Report_2019.pdf; Stern, *Why Are We Waiting?*, page 172.
38. See "Sweden Just Published a Pamphlet Preparing Its Citizens for War. What Has Got Stockholm Worried?," *Agence France-Presse*, May 22, 2018, at https://www.scmp.com/news/world/europe/article/2147229/sweden-puts-out-pamphlet-prepare-its-citizens-war-russia-tensions; Peter Apps, "Commentary: Why Neutral, Peaceful Sweden Is Preparing for War," Reuters, May 30, 2018, at https://www.reuters.com/article/us-apps-sweden-commentary/commentary-why-neutral-peaceful-sweden-is-preparing-for-war-idUSKCN1IV27N.
39. See Jay Gulledge, "Three Plausible Scenarios of Future Climate Change," in *Climate Cataclysm: The Foreign Policy and National Security Implications of Climate Change*, ed. Kurt M. Campbell (Brookings Institution Press, Washington DC 2008).
40. See General Ronald Keys (ret.) et al., "Sea Level Rise and the US Military's Mission," *The Center for Climate and Security*, 2nd ed., February 2018, at https://climateandsecurity.files.wordpress.com/2018/02/military-expert-panel-report_sea-level-rise-and-the-us-militarys-mission_2nd-edition_02_2018.pdf.
41. For the full text, see https://www.whitehouse.gov/wp-content/uploads/2017/12/NSS-Final-12-18-2017-0905.pdf.
42. For the structure and tasks of the Corps of Engineers, see https://www.usa.gov/federal-agencies/u-s-army-corps-of-engineers/.
43. See, for example, Arno J. Mayer, *The Persistence of the Old Regime: Europe to the Great War* (Pantheon Books, New York 1981), pages 3–16, 79–128.
44. American Security Project at https://www.americansecurityproject.org/; Center for Climate and Security at https://www.americansecurityproject.org/.

45. John Kerry et al., "Letter to the President of the United States: 58 Senior Military and National Security Leaders Denounce NSC Climate Panel," March 5, 2019, at https://climateandsecurity.org/letter-to-the-president-of-the-united-states-nsc-climate-panel/.
46. See Seth Jacobson, "Britain's Top General Warns of Existential Threats to Nation," *The National,* December 12, 2018, at https://www.thenational.ae/world/europe/britain-s-top-general-warns-of-existential-threats-to-nation-1.801739; General Mark Carleton-Smith, "Russia Poses Greater Threat Than ISIS, New British Army Chief Warns," *The Guardian*, November 24, 2018; Air Chief Marshal Sir Stuart Peach, "Valedictory Address as Chief of the Defense Staff," June 5, 2018, at https://policyexchange.org.uk/wp-content/uploads/2018/06/CDS-transcript.pdf; Thomas Mackie, "World War 3 Fears: Military Expert Warns UK Is at 'Real Risk of War' with Russia and Putin," *Daily Express*, March 17, 2018, at https://www.express.co.uk/news/world/933068/world-war-3-nuclear-war-russia-vladimir-putin-admiral-lord-west.
47. See Ghosh, *Great Derangement*, pages 138–140.
48. See Esther Babson, "Climate Change Impacts on National Security," *American Security Project* fact sheet, February 2019, at https://www.americansecurityproject.org/wp-content/uploads/2019/05/Climate-Change-Impacts-on-Natsec.pdf.
49. US Department of Defense, Quadrennial Defense Review (QDR) 2018, at https://dod.defense.gov/Portals/1/Documents/pubs/2018-National-Defense-Strategy. For the exaggeration of security threats by the bipartisan Washington establishment, see Stephen M. Walt, *The Hell of Good Intentions: America's Foreign Policy Elite and the Decline of U.S. Primacy* (Farrar, Straus and Giroux, New York 2018), pages 137–172.
50. Chad Michael Briggs, "Climate Security, Risk Assessment and Military Planning," *International Affairs*, vol. 88, no. 5, 2012.
51. Tyler H. Lippert, "NATO, Climate Change and International Security: A Risk Governance Approach," https://www.rand.org/pubs/rgs_dissertations/RGSD387.html.
52. UK Cabinet Office 2008: 18, at https://assets.publishing.service.gov.uk/government/uploads/system/uploads/attachment_data/file/229001/7590.pdf.
53. Quoted in Hulme, *Why We Disagree*, page 284.
54. Heather A. Conley and Matthew Melino, "The Implications of US Policy Stagnation toward the Arctic Region," Centre for Strategic and International Studies, Washington DC, May 3, 2019, at https://www.csis.org/analysis/implications-us-policy-stagnation-toward-arctic-region. See also Mathieu Boulegue, "NATO Needs a Strategy for Countering Russia in the Arctic and the Black Sea," Royal Institute of International Affairs (Chatham House), July 2, 2018, at https://www.chathamhouse.org/expert/comment/nato-needs-strategy-countering-russia-arctic-and-black-sea;

Shiloh Rainwater, "Race to the North: China's Arctic Strategy and Its Implications," *US Naval War College Review*, vol. 66, no. 2, Spring 2013; Anastasia Astrashevskaya and Henry Foy, "Polar Powers: Russia's Bid for Supremacy in the Arctic Ocean," *Financial Times*, April 28, 2019, at https://www.ft.com/content/2fa82760-5c4a-11e9-939a-341f5ada9d40.

55. Adam Vaughn, "UK Green Energy Investment Halves after Policy Changes," *The Guardian*, January 16, 2018, at https://www.theguardian.com/business/2018/jan/16/uk-green-energy-investment-plunges-after-policy-changes; official figures for military spending from the UK Ministry of Defence at assets.publishing.service.gov.uk/government/uploads/system/uploads/attachment_data/file/494526/FOI2015-08279-Cost_of_the_wars_in_Iraq_and_Afghanistan.pdf.
56. R. James Woolsey, "Security Implications of Climate Scenario," in Kurt M. Campbell et al., *The Age of Consequences: The Foreign Policy National Security Implications of Global Climate Change* (Center for Strategic and International Studies, Washington DC, 2007), pages 85–86.
57. Simon Dalby, *Security and Environmental Change* (Polity Press, Cambridge UK 2009), pages 155–156.
58. Jamie Smyth, "Floods and Fires Prompt Canberra Crisis Talks," *Financial Times*, February 9–10, 2019.
59. See also the papal *Encyclica Laudato Si* (2015), page 44, at https://laudatosi.com.
60. For a good way of approaching the issue of risk with regard to climate change, see Guzman, *Overheated*, page 87.
61. HSBC, Hongkong and Shanghai Banking Corporation. Ashim Paun, Lucy Acton, and Wai-Shin Chan, "Fragile Planet: Scoring Climate Risks around the World," HSBC Global Research, March 2018, at https://www.sustainablefinance.hsbc.com/reports/fragile-planet.
62. "Climate Change in Central America: Potential Impacts and Public Policy Options," *Economic Commission for Latin America and the Caribbean* (*ECLAC*), (UN Publications 2018), at https://repositorio.cepal.org/bitstream/handle/11362/39150/7/S1800827_en.pdf; "How Is Climate Change Affecting Mexico?," Climate Reality Project, February 15, 2018, at https://www.climaterealityproject.org/blog/how-climate-change-affecting-mexico; James G. Lamb, "Climate Change: Existential Threats in a Time of Denial," *Berkeley Review of Latin American Studies*, Fall 2017.
63. Kerry Emanuel, "The Hurricane-Climate Connection," *American Meteorological Society*, May 2008.
64. For the connection between climate change and Central American migration, see Rachel Hagen, "Desperation and Drought: Why Thousands Flee," *American Security Project*, November 8, 2018, at https://www.americansecurityproject.org/desperation-and-drought-why-thousands-flee; Adam Wernick, "Climate Change Is the Overlooked Driver of Central American Migration," Public Radio International,

February 6, 2019, at https://www.pri.org/stories/2019-02-06/climate-change-overlooked-driver-central-american-migration.

65. Adam Tooze, "Rising Tides Will Sink Global Order," *Foreign Policy*, December 20, 2018.
66. "Climate Change Impacts on Human Health," US Environmental Protection Agency, January 2017, at https://19january2017snapshot.epa.gov/climate-impacts/climate-impacts-human-health_.html; "The Impacts of Climate Change on Human Health in the United States: A Scientific Assessment," US Global Change Research Program 2016, at https://health2016.globalchange.gov/; Elizabeth G. Hanna, "Health Hazards," in *The Oxford Handbook of Climate Change and Society*, 217–231; Paul R. Epstein, "Climate and Health," *Science*, vol. 285, no. 5426, July 16, 1999, pages 347–348; Jonathan A. Patz et al., "The Effects of Changing Weather on Public Health," *Annual Review of Public Health*, vol. 21, 2000, pages 271–307, at https://www.annualreviews.org/abs/doi/10.1146/annurev.publhealth.21.1.271?intcmp=trendmd.
67. See the report by the US Environmental Protection Agency of January 19, 2017 (before it was crippled by Trump), "Climate Impacts on Coastal Areas," at https://19january2017snapshot.epa.gov/climate-impacts/climate-impacts-coastal-areas_.html; for some particular case studies of particular places from which people will have to move, see "Climigration Case Studies" at http://www.climigration.org/case-studies.
68. See "Underwater: Rising Seas, Chronic Floods, and the Implications for US Coastal Real Estate," Union of Concerned Scientists, 2018, at https://www.ucsusa.org/global-warming/global-warming-impacts/sea-level-rise-chronic-floods-and-us-coastal-real-estate-implications#.WyqrH9Izrcs. See also Debra Javeline and Tracy Kijewski-Correa, "Coastal Homeowners in a Changing Climate," *Climatic Change*, vol. 152, no. 2, 2018, pages 259–274.
69. See Amos Tai et al., "Threat to Future Global Food Security from Climate Change," *Nature Climate Change*, vol. 4, 2014, pages 817–821.
70. See Kendra Pierre-Louis, "Want to Escape Global Warming? These Cities Promise Cool Relief," *New York Times*, April 15, 2019.
71. Russian Public Opinion Research Center (VCIOM), June 2014, at https://www.wciom.com/.
72. L. A. Henry and L. M. Sundstrom, "Russia's Climate Policy: International Bargaining and Domestic Modernisation," *Europe-Asia Studies*, vol. 64, no. 7, 2012, pages 1297–1322.
73. Renat Perelet, Sergei Pegov, and Mighail Yukgin, "Climate Change: Russia Country Paper," United Nations Development Programme Human Development Report 2008, at http://hdr.undp.org/en/content/climate-change-russia-country-paper; Anna Korppoo, "Russia and Climate Change: Costs or Benefits?," *The World Today*, May 2009, at http://www.upi-fiia.fi/assets/wt050908.pdf.

74. Niels Smeets, "Combating or Cultivating Climate Change? Russia's Approach to Global Warming as an Obstacle to EU Environmental Pioneering," *Academic Association for Contemporary European Studies*, at https://www.uaces.org/archive/papers/abstract.php?paper_id=834.
75. Steven Solomon, *Water: The Epic Struggle for Wealth, Power and Civilization* (Harper Collins, New York 2010), pages 430–447.
76. Charlie Parton, "China's Looming Water Crisis," *China Dialogue,* April 2018, page 6, at https://chinadialogue-production.s3.amazonaws.com/uploads/content/file_en/10608/China_s_looming_water_crisis_v.2__1_.pdf; Ben Abbs, "The Growing Water Crisis in China," *Global Risks Insights*, August 10, 2017, at https://globalriskinsights.com/2017/08/shocks-china-growing-water-crisis/; "China: A Watershed Moment for Water Governance," World Bank, November 7, 2018, at https://www.worldbank.org/en/news/press-release/2018/11/07/china-a-watershed-moment-for-water-governance.
77. Ines Peres, "Climate Change, Drought and Rising Food Prices Heightened Arab Spring," *Scientific American*, March 4, 2013, at https://www.scientificamerican.com/article/climate-change-and-rising-food-prices-heightened-arab-spring/?redirect=1; Troy Sternberg, "Chinese Drought, Bread and the Arab Spring," *Applied Geography*, vol. 34, no. 4, May 2012, pages 519–524.
78. For growing water scarcity in Afghanistan, see Soraya Parwani, "Is Water Scarcity a Bigger Threat to Afghanistan than the Taliban?," *The Diplomat*, October 10, 2018, at https://thediplomat.com/2018/10/is-water-scarcity-a-bigger-threat-than-the-taliban-in-afghanistan/.
79. Guzman, *Overheated,* pages 90–91.
80. For a classic study of Chinese banditry and its origins in the first half of the 20th century, see Phil Billingsley, *Bandits in Republican China* (Stanford University Press, Palo Alto, CA, 1988).
81. For the Manchu dynasty's loss of control of the Yellow River, see David A. Pietz, *The Yellow River: The Problem of Water in Modern China* (Harvard University Press, Cambridge, MA 2015), pages 64–69. For subsequent strategies of controlling it, see pages 79–193, 259–321.
82. For the founding document of China's climate change strategy, see "China's National Climate Change Programme," National Development and Reform Commission, 2007, at http://www.china-un.org/eng//gyzg/t626117.htm.
83. Scott Melton and Michelle Melton, "China's Pivot on Climate Change and National Security," *Lawfare*, April 2, 2019, at https://www.lawfareblog.com/chinas-pivot-climate-change-and-national-security.
84. "How Is China Managing Its Greenhouse Gas Emissions?," *China Power,* March 2019, at https://chinapower.csis.org/china-greenhouse-gas-emissions/.

85. Stephen Chen, "China Cracks Cheap Lithium Production in Electric Car Breakthrough," *South China Morning Post,* May 14, 2019; Amanda Lee, "China's Electric Car Market Is Growing Twice as Fast as US. Here's Why," *South China Morning Post,* April 27, 2018; Isabelle Niu, "The Future of Electric Cars Is Happening Now in China," *Quartz,* April 8, 2019, at https://qz.com/1586938/china-will-dominate-the-worlds-electric-cars-market/.
86. Caleb Mills, "Water Shortages: China's Unrecognised Threat," *Geopolitical Monitor,* June 6, 2018, at https://www.geopoliticalmonitor.com/water-shortages-chinas-unrecognized-threat/; Matthew Carney, "Forget Geopolitics. Water Scarcity Shapes Up as the Biggest Threat to China's Rise," ABC Australia, November 26, 2018, at https://www.abc.net.au/news/2018-11-23/china-water-crisis-threatens-growth/10434116.
87. For the indifference of South Asian publics to climate change, see Ghosh, *Great Derangement*, pages 125–126.
88. See the World Food Programme's Food Security Analyses at http://www.wfp.org/climate-change/innovations/analyses; Jeremy S. Pal and Elfatih A. Eltahir, "Future Temperature in Southwest Asia Projected to Exceed a Threshold for Human Adaptability," *Nature Climate Change,* vol. 6, no. 2, 2016, pages 197–200.
89. Ashim Paun, Lucy Acton and Wai-Shin Chan, "Fragile Planet: Scoring Climate Risks around the World," HSBC Global Research, March 2018, at https://www.sustainablefinance.hsbc.com/reports/fragile-planet.
90. Esther Babson, "The Importance of Rice: Why We Should Care about Sri Lanka's Changing Climate," American Security Project, January 8, 2018, at https://www.americansecurityproject.org/the-importance-of-rice.
91. "Global Warming of 1.5 Degrees", Intergovernmental Panel on Climate Change (IPCC), 2018, at https://report.ipcc.ch/sr15/pdf/sr15_spm_final.pdf.
92. For the threat of climate change to already fragile states, see Caitlin E. Werrell and Francesco Femia, "Climate Change, the Erosion of State Sovereignty and World Order," *Brown Journal of World Affairs*, vol. 12, issue 2, Spring–Summer 2016, pages 221–235, at http://bjwa.brown.edu/22-2/climate-change-the-erosion-of-state-sovereignty-and-world-order/; "Fifth IPCC Assessment Report: What's in It for South Asia?," Intergovernmental Panel on Climate Change, 2014, at https://cdkn.org/wp-content/uploads/2014/04/CDKN-IPCC-Whats-in-it-for-South-Asia-AR5.pdf; Tariq Waseem Ghazi, A. N. M. Munirazzaman and A. K. Singh, "Climate Change and Security in South Asia: Co-operating for Peace," Global Military Council on Climate Change (GMACCC), May 2016, at http://gmaccc.org/wp-content/uploads/2016/05/Climate_Change_and_Security_in_South_Asia.pdf; "Climate Change and Risks to Food Security," World Economic Forum Global Risks Report 2016 (with Marsh

and McLennan Companies and Zurich Insurance Group), pages 50–58, at https://www.weforum.org/reports/the-global-risks-report-2016.

93. Muthukumara Mani, Sushenjit, Bandyopadhyay, Shun, Chonabayashi, Anil Markandya and Thomas Mosier, *South Asia's Hotspots: Impact of Temperature and Precipitation Changes on Living Standards* (World Bank, Washington DC June 2018).
94. Andy Turner, "The Indian Monsoon in a Changing Climate," Royal Meteorological Society, London, *Composite Water Management Index*, August 16, 2018, at https://www.rmets.org/resource/indian-monsoon-changing-climate; National Institute for Transforming India (NITI), June 2018, at www.niti.gov.in; David Antos, "India, Climate Change and Security in South Asia," *Center for Climate and Security* briefer no. 36, May 3, 2017.
95. Soutik Biswas, BBC, June 10, 2019, at https://www.bbc.com/news/world-asia-india-48552199.
96. Christian Parenti, *Tropic of Chaos: Climate Change and the New Geography of Violence* (Nation Books, New York 2011), pages 133–156.
97. "Climate Change Could Depress Living Stanadrds in India", The World Bank, June 28, 2018, at https://www.worldbank.org/en/news/press-release/2018/06/28/climate-change-depress-living-standards-india-says-new-world-bank-report
98. "Composite Water Management Index," June 2018, at http://www.niti.gov.in/writereaddata/files/document_publication/2018-05-18-Water-index-Report_vS6B.pdf.
99. Avinash Mishra and Devashish Dhar, "India Needs to Focus on Water Efficiency," *LiveMint*, July 12, 2018; Solomon, *Water*, pages 421–422.
100. For the growing water crisis in Bangalore, see Carly Cassella, "A Major Indian City Has Almost Run Out of Water, and No One Is Talking about It," *Science Alert*, June 21, 2019, at https://www.sciencealert.com/this-indian-city-has-all-but-run-out-of-water-and-no-one-is-talking-about-it; Sowmya Rajan and Barkha Kumari, "2020: The Year Bangaluru Runs Out of Water," *Bangalore Mirror*, June 29, 2019, at https://bangaloremirror.indiatimes.com/bangalore/cover-story/2020-the-year-bengaluru-runs-out-of-water/articleshow/69995755.cms.
101. Quoted in David Hazony, "How Israel Is Solving the Global Water Crisis," *The Tower*, vol. 31, October 2015, at http://www.thetower.org/article/how-israel-is-solving-the-global-water-crisis/.
102. Parton, "China's Looming Water Crisis," pages 19–20.
103. For the disastrous 2005 floods in Mumbai due to absence of planning for floods, see R. B. Bhagat, Mohua Guha, and Aparajita Chattopadhyay, "Mumbai after 26/7 Deluge: Issues and Concerns in Urban Planning," *Population and Environment* no. 27, 2006, pages 337–349. For a portrait of the 2005 floods and a dire warning of the consequences if a major cyclone were to strike the city, see Ghosh, *Great Derangement*, pages 45–54.

104. Guzman, *Overheated*, pages 54–71; Tariq Waseem Ghazi et al., "Climate Change and Security in South Asia: Co-operating for Peace."
105. Quoted in Parenti, *Tropic of Chaos*, page 139.

Chapter 2

1. Paul Collier, *Exodus: How Migration Is Changing Our World* (Oxford University Press, New York 2015), page 30.
2. See the International Institute for Strategic Studies, "The IISS Transatlantic Dialogue on Climate Change and Security: Report to the European Commission," 2011, at http://www.andrew-holland.com/uploads/6/3/1/7/6317360/climate_change__security_final_report.pdf.
3. See the International Institute for Strategic Studies, "The IISS Transatlantic Dialogue on Climate Change and Security: Report to the European Commission," 2011, at http://www.andrew-holland.com/uploads/6/3/1/7/6317360/climate_change__security_final_report.pdf.
4. See *Foresight: Migration and Global Environmental Change: Final Project Report*, Government Office for Science, London, 2011, pages 11–12 at https://assets.publishing.service.gov.uk/government/uploads/system/uploads/attachment_data/file/287717/11-1116-migration-and-global-environmental-change.pdf.
5. https://www.nato.int/docu/Review/2010/; Sharon Burke, "Catastrophic Climate Change over the Next Hundred Years," in *Climate Cataclysm: The Foreign Policy and National Security Implications of Climate Change*, ed. Kurt M. Campbell (Brookings Institution Press, Washington DC 2008), page 164; Mark Lynas, *Six Degrees: Our Future on a Hotter Planet* (Fourth Estate, London 2007), page 141.
6. "Long-Standing Partisan Gap over Views of Compromise Disappears," *Pew Research Center, Survey of US Adults*, March 7, 2018, at https://www.people-press.org/2018/04/26/8-the-tone-of-political-debate-compromise-with-political-opponents/8_6/; Jeremiah J. Castle et al., "Survey Experiments on Candidate Religiosity, Political Attitudes, and Vote Choice," *Journal for the Scientific Study of Religion*, April 6, 2017, at https://doi.org/10.1111/jssr.12311; Peter Beinart, "Secular Democrats Are the New Normal," *The Atlantic*, March 15, 2019.
7. The Somali population in England—strongly present in my part of London—are a particularly dire example of people bringing the social dysfunction of their home society with them. See Ismail Einashe, "Mo and Me," *Prospect*, September 2012.
8. Eric Kaufman, *Whiteshift* (Abrams Press, New York 2019).
9. For speculation along these lines, see Yuval Noah Harari, *Homo Deus: A Brief History of Tomorrow* (Harper, New York 2018); and Francis Fukuyama, *Our Posthuman Future: Consequences of the Biotechnological Revolution* (Picador, New York 2003).

10. See Michael Hardt and Antonio Negri, *Empire* (Harvard University Press, Cambridge MA 2000), pages 397–413; Alain Badiou, *Theory of the Subject*, translated by Brunor Bosteels (1982, reprinted Bloomsbury, London 2013); Alain Badiou, "True Communism Is the Foreignness of Tomorrow," talk in Athens, January 25, 2014, at https://www.versobooks.com/blogs/1547-true-communism-is-the-foreignness-of-tomorrow-alain-badiou-talks-in-athens; Thomas Nail, "The Political Centrality of the Migrant," in *Critical Perspectives on Migration in the 21st Century*, ed. Marianna Karakoulaki, Laura Southgate, and Jakob Steiner (E-International Relations Publishing, London 2018).
11. For environmentalist arguments for the undermining of the nation state, see, for example, Paul G. Harris, *Global Ethics and Climate Change* (Edinburgh University Press, Edinburgh 2016); Robert Muggah et al., "Cities, Not Nation States Will Determine Our Future Survival," World Economic Forum, June 2, 2017, at https://www.weforum.org/agenda/2017/06/as-nation-states-falter-cities-are-stepping-up/; Naomi Klein, *This Changes Everything* (Simon and Schuster, New York 2014); and the chapters by Sheila Jasanoff, Harriet Bulkeley, Ronnie D. Lipschutz, Karin Backstrand, Frank Bierman, and others in *The Oxford Handbook of Climate Change and Society*, ed. John S. Dryzek, Richard B. Norgaard, and David Schlosbert (Oxford University Press, New York 2011), at https://www.weforum.org/agenda/2017/06/as-nation-states-falter-cities-are-stepping-up/andbook.
12. See Etienne Piguet, Antoine Pecoud, and Paul de Guchteneire, eds., *Migration and Climate Change* (Cambridge University Press and UNESCO, Cambridge 2011), pages 3–4; Jennifer Gordon, "People Are Not Bananas: How Immigration Differs from Trade," *Northwestern University Law Review*, vol. 104. no. 3, 2010. The fact that it is apparently necessary to inform many economists that *Homo sapiens* is not a species of tropical fruit is a truly striking comment on the economics profession. For a standard view from the international capitalist establishment, see Ian Goldin, "How Immigration Has Changed the World—For the Better," World Economic Forum report, January 17, 2016, at https://www.weforum.org/agenda/2016/01; Frans Timmermans, vice president of the European Commission, speech to the European Parliament, "Diversity Is Humanity's Destiny," at https://www.youtube.com/watch?v=YMeQKI-VprQ.
13. Christian Parenti, *Tropic of Chaos: Climate Change and the New Geography of Violence* (Nation Books, New York 2011), pages 179–224.
14. For the taboos on discussing migration, see Collier, *Exodus*, pages 22–26, 61–62; David Goodhart, *The British Dream: Successes and Failures of Post-War Immigration* (Atlantic Books, London 2013), pages xx–xxvii.
15. See, for example, "World Publics Welcome Global Trade—but Not Immigration," Pew Research Center, October 4, 2007; "Standard

Eurobarometer 89" (Brussels: European Commission, June 2018), at http://ec.europa.eu/commfrontoffice/publicopinion/index.cfm/Survey/getSurveyDetail/instruments/STANDARD/surveyKy/2180. For the 2010 elections in Britain, see Geoffrey Evans and Yekaterina Chzhen, "Explaining Voters' Defection from Labour over the 2005–2010 Electoral Cycle: Leadership, Economics and the Rising Importance of Immigration," *Political Studies,* April 2013. For the sources of the rise of the populist right, see William A. Galston, *Anti-Pluralism: The Populist Threat to Liberal Democracy* (Yale University Press, New Haven CT 2018); Ivan Krastev, "3 Versions of Europe Are Collapsing at the Same Time," *Foreign Policy,* July 10, 2018, at https://foreignpolicy.com/2018/07/10/3-versions-of-europe-are-collapsing-at-the-same-time/.; James Kirchick, "Centre Right Strategies for Addressing the Rise of the European Far Right," *Brookings Institution,* policy brief, at https://www.brookings.edu/wp-content/uploads/2019/02/FP_20190226_european_far_right_kirchick.pdf; Marc F. Plattner, "Illiberal Democracy and the Struggle on the Right," *Journal of Democracy,* vol. 30, no. 1, January 2019, pages 15–19, at https://www.journalofdemocracy.org/articles/illiberal-democracy-and-the-struggle-on-the-right/.

16. International Office of Migration, *World Migration Report 2018*, at http://publications.iom.int/books/world-migration-report-2018; Jonathan L. Bamber et al., "Ice Sheet Contributions to Future Sea Level Rise from Structured Expert Judgment," *Proceedings of the National Academy of Sciences USA,* May 20, 2019, at https://www.pnas.org/content/early/2019/05/14/1817205116.
17. Timothy Doyle and Sanjay Chaturvedi, "Climate Refugees and Security: Conceptualisations, Categories and Contestations," in *The Oxford Handbook of Climate Change and Society,* page 288.
18. Barack Obama, "A More Perfect Union," speech in Philadelphia, March 18, 2008, quoted in Paul Scheffer, *Immigrant Nations* (Polity Press, Cambridge 2011), pages 21–22.
19. For Tucker Carlson's remarks, see https://www.foxnews.com/opinion/tucker-carlson-mitt-romney-supports-the-status-quo-but-for-everyone-else-its-infuriating; Mark Lilla, *The Once and Future Liberal: After Identity Politics* (Harper, New York 2017); Thomas Frank, *Listen, Liberal: Or, Whatever Happened to the Party of the People* (Harper, New York 2017).
20. Paul Collier, *The Future of Capitalism* (Allen Lane, London 2018), pages 194–198.
21. Thomas Homer-Dixon and Jessica Blitt, eds., *Ecoviolence: Links among Environment, Population and Security* (Rowman and Littlefield, Lanham MD 1998); Thomas Homer-Dixon, *Environment, Scarcity and Violence* (Princeton University Press, Princeton NJ 1998).
22. Carolina Fritz, "Climate Change and Migration: Sorting through Complex Issues without the Hype," Migration Policy Institute (MPI), March 4, 2010, at https://www.migrationpolicy.org/article/

climate-change-and-migration-sorting-through-complex-issues-without-hype; Philip Nel and Marjolein Righarts, "Natural Disasters and the Risk of Violent Civil Conflict," *International Studies Quarterly*, no. 52, 2008, pages 159–185; Martine Rebetez, "The Main Climate Change Forecasts That Might Cause Human Displacements," in Piguet et al., *Migration and Climate Change*, pages 37–48; Robert Stojanov, "Environmental Migration: How Can It Be Estimated and Predicted?" *Geographica*, vol. 38, 2004, pages 77–84; Sabine Perch-Nielsen, Michele Battig, and Dieter Imboden, "Exploring the Link between Climate Change and Migration," *Climatic Change*, vol. 91, 2004, pages 375–393.

23. See Kevin Whitelaw, "Climate Change Will Have Destabilizing Consequences, Intelligence Agencies Warn," *US News and World Report*, June 25, 2008.
24. See Peter H. Gleick, "Water, Drought, Climate Change and Conflict in Syria," *Weather, Climate and Society*, July 1, 2014.
25. See Eleanor J. Burke, Simon J. Brown, and Nikolaos Christidis, "Modelling the Recent Evolution of Global Drought and Projections for the 21st Century with the Hadley Centre Climate Model," *Journal of Hydrometeorology*, vol. 7, no. 5, October 2006, pages 1113–1125.
26. See Christian Parenti, *Tropic of Chaos: Climate Change and the New Geography of Violence* (Nation Books, New York 2011), pages 39–96.
27. According to the International Crisis Group, "The conflict's roots lie in the—often forced—migration of herders south from their traditional grazing grounds in northern Nigeria. As drought and desertification have dried up springs and streams across Nigeria's far northern Sahelian belt, large no.s of herders have had to search for alternative pastures and sources of water for their cattle." *Stopping Nigeria's Spiralling Farmer-Herder Violence*, International Crisis Group, report no. 262/Africa, 26 July 2018, at https://www.crisisgroup.org/africa/west-africa/nigeria/262-stopping-nigerias-spiralling-farmer-herder-violence.
28. Estimates at http://www.worldometers.info/world-population/africa-population/.
29. *Foresight: Migration and Global Environmental Change*, page 13.
30. Kanti Kumari Rigaud et al., "Groundswell: Preparing for Internal Climate Migration," World Bank Group 2018, at https://openknowledge.worldbank.org/handle/10986/29461.
31. See Luigi Scazzieri, "Tearing at Europe's Core: Why France and Italy Are at Loggerheads," Centre for European Reform, February 12, 2019; Ilaria Maria Sala, "Why Is Italy Picking a Fight with France?," *New York Times*, February 11, 2019.
32. Myron Weiner, *The Global Migration Crisis: Challenge to States and Human Rights* (Harper Collins, London 1995), page 88.
33. Myron Weiner, *Sons of the Soil: Migration and Ethnic Conflict in India* (Princeton University Press, Princeton NJ 1978), pages 3–18; Milton

Esman, "Communal Conflict in Southeast Asia," in *Ethnicity: Theory and Experience*, ed. Nathan Glazer and Patrick Moynihan (Harvard University Press, Cambridge MA 1975), pages 391–416.

34. Weiner, *Sons of the Soil*, page 307.
35. Weiner, *Sons of the Soil*, pages 308–309.
36. Laurent Gayer, *Karachi: Ordered Disorder and the Struggle for the City* (C. Hurst, London 2014); Anatol Lieven, *Pakistan: A Hard Country* (Penguin, London 2011), pages 309–338; Oskar Verkaaik, *Migrants and Militants: Fun and Urban Violence in Pakistan* (Princeton University Press, Princeton, NJ 2004); Hastings Donnan and Pnina Werbner, *Economy and Culture in Pakistan: Migrants and Cities in a Muslim Society* (Macmillan, London 1991).
37. David Antos, "India, Climate Change and Security in South Asia," Center for Climate and Security, Washington, DC, May 3, 2017, at https://climateandsecurity.files.wordpress.com/2012/04/india_climate-change-and-security-in-south-asia_briefer-36.pdf.
38. Eleanor Albert and Andrew Chatzky, "The Rohingya Crisis," Council on Foreign Relations, New York, December 2018, at https://www.cfr.org/backgrounder/rohingya-crisis.
39. Courtney Subramanian, "India Election 2019: Echoes of Trump in India's Border Politics," BBC, May 21, 2019, at https://www.bbc.com/news/world-asia-india-48334689.
40. "Statistics on Violations in the Border Area," Odhikar, November 2018, at http://odhikar.org/statistics/statistics-on-violations-in-the-border-area/; Kamrul Hasan, "Diplomacy Drives Down Border Deaths," *Dhaka Tribune*, September 9, 2018, at https://www.dhakatribune.com/bangladesh/foreign-affairs/2018/09/09/diplomacy-drives-down-border-deaths.
41. Sanjeev Tripathi, "Illegal Migration from Bangladesh to India: Towards a Comprehensive Solution," Carnegie India, July 29, 2016, at https://carnegieindia.org/2016/06/29/illegal-immigration-from-bangladesh-to-india-toward-comprehensive-solution-pub-63931. See also Allan Findlay and Alistair Geddes, "Critical Views on the Relationship between Climate Change and Migration: Some Insights from the Experience of Bangladesh," in *Migration and Climate Change*, pages 138–159.
42. Kanta Kumari Rigaud et al., "Groundswell: Preparing for Internal Climate Migration," World Bank 2018, page 146, at https://openknowledge.worldbank.org/handle/10986/11866.
43. Emil Pain, "Xenophobia and Ethnopolitical Extremism in Post-Soviet Russia: Dynamics and Growth Factors," *Nationalities Papers*, vol. 35, no. 5, 2007, pages 895–911; Anastasia Gorodeisky et al., "The Nature of Anti-Immigrant Sentiment in Post-Socialist Russia," *Post-Soviet Affairs*, vol. 31, no. 2, 2015, pages 115–135; Alexey Timofeychev, "Is Russia on the Brink of a Migration Crisis?," *Russia Beyond*, February 18, 2019, at https://

www.rbth.com/lifestyle/329990-has-russia-migration-crisis; James Brooke, "Russian Nationalists March to an Anti-Immigrant Drum," Voice of America, November 4, 2013, at https://www.voanews.com/a/russia-nationalists-march-to-an-anti-immigration-drum/1785223.html.

44. Matthew Cooper and Bradley Jardine, "For Russia's Labor Migrants, a Life on the Edge," *Moscow Times,* November 4, 2016, at https://www.themoscowtimes.com/2016/11/04/for-labor-migrants-a-life-on-the-edge-a56018.
45. See "Alexey Navalny on Putin's Russia" (interview with Shaun Walker), *The Guardian*, April 29, 2017, at https://www.theguardian.com/world/2017/apr/29/alexei-navalny-on-putins-russia-all-autocratic-regimes-come-to-an-end.
46. Thomas Grove, "Thousands of Russian Nationalists Rally in Anti-Immigrant Protests," Reuters, November 4, 2013, at https://www.reuters.com/article/us-russia-nationalists/thousands-of-russian-nationalists-rally-in-anti-immigrant-protest.
47. See Raymond Sontag, "Putin's Nationalist Gamble," *American Interest,* March 14, 2014, at https://www.the-american-interest.com/2014/03/14/putins-nationalist-gamble/.
48. Vladimir Putin, "Russia: The Ethnicity Issue," *Nezavisimaya Gazeta*, January 23, 2012, at http://archive.premier.gov.ru/eng/events/news/17831/ . See also President Vladimir Putin, State of the Union Address 2012, at https://eng.kremlin.ru/transcripts/4739.
49. For Russian official policy on migration and assimilation as of 2019, see "Decree on the Concept of the State Migration Policy of the Russian Federation for 2019–2025," October 21, 2018, at http://kremlin.ru/events/president/news/58986.
50. "Amid Ethnic Unrest, Russian Governor Bans Migrant Workers from Dozens of Jobs in Construction, Farming and Public Transport," *Meduza*, March 28, 2019, at https://meduza.io/en/news/2019/03/28/.
51. Sener Akturk, "How Immigration Aids Russia's Transformation into an Assimilationist Nation-State," PONARS policy memo no. 347, September 2014, at http://www.ponarseurasia.org/memo/how-immigration-aids-russias-transformation-assimilationist.
52. For the danger that the idea of Russia as a nation state poses to Russia's ethnic minorities, see Anatol Lieven, *Chechnya: Tombstone of Russian Power* (Yale University Press, New Haven CT 1999), pages 381–384.
53. Dani Rodrik, *The Globalisation Paradox: Democracy and the Future of the World Economy* (W. W. Norton, New York 2011), page xix.
54. For the Green New Deal proposal of 2018–19, see https://www.dataforprogress.org/green-new-deal. See also Demetri Sevastopolu and Courtney Weaver, "Leftwingers Fire Up the Democrats," *Financial Times,* February 16–17, 2019; Jedediah Britton-Purdey, "The Future of the Planet Meets Policy," *New York Times,* February 16–17, 2019.

55. For a longer exposition of my views on migration to Europe, see my debate with Philippe Fargues in *Debating Migration to Europe: Welfare vs. Identity*, ed. Raffaele Marchetti (Routledge, London 2018).
56. For an argument along these lines as far as France is concerned, see Gerard Noirel, *Le Creuset Francais: Histoire de l'immigration xixe–xxe Siecle* (Editions du Seuil, Paris 1988).
57. See, for example, Philippe van Parijs and Yannick Vanderborght, *Basic Income: A Radical Proposal for a Free Society and a Sane Economy* (Harvard University Press, Cambridge MA 2017); Rutger Bregman, *Utopia for Realists: How We Can Build the Ideal World* (Little, Brown, London 2017); Ian Bremmer, *Us vs. Them: The Failure of Globalism* (Penguin, New York 2018), pages 147–149, discusses the old roots of the UBI idea; Robert Reich, *Saving Capitalism: For the Many, Not the Few* (Vintage, New York 2016), pages 213–217.
58. For a successful small-scale experiment in Canada, see Bryan Bergstein, "Basic Income Could Work—If You Do It Canada-Style," *MIT Technology Review*, June 21, 2018, at https://medium.com/mit-technology-review/basic-income-could-work-if-you-do-it-canada-style-2d483e733165.
59. David Goodhart, *The Road to Somewhere: The New Tribes Shaping British Politics* (Penguin, London 2017), pages 120–121; Scheffer, *Immigrant Nations*, page 81.
60. See the UK Office of Budgetary Responsibility estimate that immigration has driven up property prices by some 10 percent, thereby contributing significantly to the housing crisis in southern England and the pricing of many indigenous people out of the housing market. Cited in Collier, *Exodus*, page 127. German figure quoted in Bremmer, *Us vs. Them*, page 23.
61. For Dubai, which has a very similar pattern, see Collier, *Exodus*, pages 149–151.
62. Edward Luce, *The Retreat of Western Liberalism* (Little, Brown, New York 2017), pages 99–102.
63. Carl Benedikt Frey and Michael A. Osborne, "The Future of Employment: How Susceptible Are Jobs to Computerisation?" (Oxford Martin Programme on Technology and Employment, Oxford University, 2013), pages 13, 19, at http://www.oxfordmartin.ox.ac.uk/downloads/academic/future-of-employment.pdf. See also P. Beaudry, D. A. Green, and B. M. Sand, "The Great Reversal in the Demand for Skill and Cognitive Tasks," Tech. Rep., NBER Working Paper No. 18901, National Bureau of Economic Research, 2013; Daron Acemoglu and Pascual Restrepo, "Robots and Jobs: Evidence from US Labor Markets," National Bureau of Economic Research Working Paper 23285, March 2017, at https://www.nber.org/papers/w23285; Ivan Krastev, *After Europe* (University of Pennsylvania Press, Philadelphia 2017), page 24; Michael Chui et al., "Where Machines Could Replace Humans and Where They

Can't—Yet," *McKinsey Quarterly*, July 2016, at https://www.mckinsey.com/business-functions/digital-mckinsey/our-insights/where-machines-could-replace-humans-and-where-they-cant-yet; David H. Autor, Frank Levy, and Richard J. Murnane, "The Skill Content of Recent Technological Change: An Empirical Exploration," *Quarterly Journal of Economics*, vol. 118, no. 4, 2003, pp. 1279–1333; see Branko Milanovic, *Global Inequality: A New Approach for the Age of Globalization* (Harvard University Press, Cambridge, MA 2018)

64. Frey and Osborne, "The Future of Employment," page 47.
65. Robert Armstrong, "Global Banks Cut 30,000 Jobs as Investment Climate Worsens," *Financial Times*, August 12, 2019.
66. See Robert B. Reich, *Saving Capitalism: For the Many, Not the Few* (Vintage Books, New York 2016), pages 203–210; Bremmer, *Us vs. Them*, pages 16–17. For the threat to jobs in the petroleum and chemical industries, see Arlie Russsell Hochschildt, *Strangers In Their Own Land: Anger and Mourning on the American Right* (New Press, New York 2018), page 320.
67. See "Forgotten in the Banlieues: Young, Diverse and Unemployed," *The Economist*, February 23, 2013; Jonathan Laurence and Justin Vaisse, "Understanding Urban Riots in France" (Brookings Institution, Washington DC, December 1, 2006), at https://www.brookings.edu/articles/understanding-urban-riots-in-france/; Karina Piser, "The Social Ladder Is Broken: Hope and Despair in the French Banlieues," *The Nation*, March 21, 2018, at https://www.thenation.com/article/the-social-ladder-is-broken-hope-and-despair-in-the-french-banlieues/; for the tendency of social despair in the banlieues to turn some young people toward Islamist extremism, see Gilles Kepel, *Terror in France: The Rise of Jihad in the West* (Princeton University Press, Princeton NJ 2017), pages 136–139.
68. The literature on the causes and consequences of economic inequality and the decline of the old working classes is of course an enormous one. See, for example, Thomas Piketty, *Capital in the 21st Century* (Harvard University Press, Cambridge MA 2014); Wolfgang Streek, *How Will Capitalism End? Essays on a Failing System* (Verso Books, London 2016); Rodrik, *The Globalisation Paradox*; Robert Putnam, *Our Kids: The American Dream in Crisis* (Simon and Schuster, New York 2016); Thomas Frank, *Listen, Liberal! Or, Whatever Happened to the Party of the People?* (Metropolitan Books, New York 2016).
69. Michael Lind, *The Next American Nation: The New Nationalism and the Fourth American Revolution* (Simon and Schuster, New York 1995), pages 209–211.
70. Michael Servoz, "AI: The Future of Work?" (European Commission report 2019), page 70, at https://ec.europa.eu/epsc/sites/epsc/files/ai-report_online-version.pdf.

71. Speaking personally.
72. See "India's Tech Giants Are Replacing Workers with Robots—Revenues Are Rising but Not Jobs" (Harvard Business School HBX, June 13, 2013), at https://scroll.in/article/809713/; Ed Gent, "Why Artificial Intelligence Could Be a Threat to India's Growth," BBC Future, May 19, 2017, at http://www.bbc.com/future/story/20170510-why-artificial intelligence-could-be-a-threat-to-indias-growth; Anirban Sen, "Artificial Intelligence Raises Indian IT Firms' Productivity; Firms Like Infosys, TCS Likely to See Lower Hiring," *Times of India,* May 29, 2016, at http://economictimes.indiatimes.com/tech/ites/artificial intelligence-raises-indian-it-industrys-productivity-firms-like-infosys-tcs-likely-to-see-lower-hiring/articleshow/52032913.cms; see Kausjik Basu, "India Hides Job Data but Truth Is Clear," *New York Times,* February 2–3, 2019.
73. *General Social Survey 2016*, NORC, University of Chicago, at http://gss.norc.org/About-The-GSS.
74. Edward Luce, *The Retreat of Western Liberalism* (Atlantic Monthly Press, New York, 2017). Thus between 2000 and 2015, the proportion of Americans defining themselves as "lower class" (as opposed to "middle class") rose from a third to a half: Frank Newport, "Fewer Americans Identify as Middle Class," Gallup, April 28, 2015, at https://news.gallup.com/poll/182918/fewer-americans-identify-middle-class-recent-years.aspx.
75. See Yasha Mounk, *The People Versus Democracy* (Harvard University Press, Cambridge MA 2018), page 180. For the important difference between "mutual respect" between different groups and "mutual regard" between people who feel they belong to the same group, and its effect on cooperation and the willingness to support others, see Collier, *Exodus*, page 68–72.
76. See Goodhart, *British Dream*, pages 261–282.
77. Alberto Alesina and Edward Glaeser, *Fighting Poverty in the US and Europe: A World of Difference* (Oxford University Press, New York 2006); Collier, *Exodus*, pages 92–94; Goodhart, *British Dream*, pages 261–282, and Goodhart, *Road to Somewhere*, pages 222–223.
78. See Robert Putnam, "E Pluribus Unum: Diversity and Community in the 21st Century," *Scandinavian Political Studies,* vol. 30, no. 2, 2007, pages 137–174.
79. See Martin Gilens, *Why Americans Hate Welfare: Race, Media and the Politics of Anti-Poverty Policy* (University of Chicago Press, Chicago 2000).
80. Lind, *The Next American Nation.*
81. Bremmer, *Us vs. Them,* page 164.
82. Kaufman, *Whiteshift*, pages 24–25, 436–451. See also Lind, pages 288–298. For a hopeful vision of future integration (after a pretty gloomy book), see Amy Chua, *Political Tribes: Group Instinct and the Fate of Nations* (Penguin, New York 2018), pages 197–210.
83. David Miller, *On Nationality* (Clarendon Press, Oxford 1995), page 25.

84. As played and directed by Warren Beatty in the move *Bulworth* (1998).
85. Daniel Defoe, "The True-Born Englishman," at https://www.poetryfoundation.org/poems/44081/the-true-born-englishman.
86. See, for example, Frans Timmermans, vice president of the European Commission, speech to the European Parliament, "Diversity Is Humanity's Destiny," at https://www.youtube.com/watch?v=YMeQKI-VprQ; Jean-Claude Juncker, president of the European Commission, State of the Union address to the European Parliament, Strasbourg, September 9, 2015, at http://europa.eu/rapid/press-release_SPEECH-15-5614_en.htm; Ian Goldin, "How Immigration Has Changed the World—For the Better," World Economic Forum report, January 17, 2016, at https://www.weforum.org/agenda/2016/01; David Pilling, "Migration Is as Old as Humanity and Should Be Welcomed," *Financial Times,* April 11, 2019, page 9.
87. This fear above all underlies Samuel Huntington's *Who Are We? The Challenges to America's National Identity* (Simon and Schuster, New York 2005).
88. See Anatol Lieven, *America Right or Wrong: An Anatomy of American Nationalism*, 2nd ed. (Oxford University Press, New York 2012), pages 16–80.
89. For an alarmist (or alarming) but deeply researched view of the implications of the growth of Muslim populations, see Christopher Caldwell, *Reflections on the Revolution in Europe* (Penguin, London 2010). For a more measured and optimistic take on the situation in France, see Jonathan Laurence and Justin Vaisse, *Integrating Islam: Political and Religious Challenges in Contemporary France* (Brookings Institution Press, Washington DC 2006).
90. Innes Bowen, *Medina in Birmingham, Najaf in Brent: Inside British Islam* (Hurst, London 2014), page 27.
91. See "Low Rate of German-Turkish Marriages Impedes Integration," at http://www.dw.com/en/low-rate-of-german-turkish-marriages-impedes-integration/a-3134365; Migration Watch UK, "Transnational Marriage and the Formation of Ghettoes," Briefing Paper 10.12, September 22, 2005. For much higher rates of intermarriage in France, see Laurence and Vaisse, *Integrating Islam*, page 44. For the gradual increase in Turkish-German intermarriage, see Olga Nottmeyer, "Wedding Bells Are Ringing: Increased Rates of Intermarriage in Germany," Migration Policy Institute, Washington, DC, October 1, 2009, at https://www.migrationpolicy.org/article/wedding-bells-are-ringing-increasing-rates-intermarriage-germany.
92. Trevor Philips, "Sleepwalking to Segregation" (2005), at www.jisscmail.ac.uk. For Muslim ghettoization in Holland, see Paul Scheffer, "The Land of Arrival," in Rene Cuperus et al., *The Challenge of Diversity: European Social Democracy Facing Migration, Integration and Multiculturalism*

(Studienverlag, Innsbruck 2003), pages 23–30. For France, see George Packer, "The Other France," *New Yorker,* August 24, 2015.

93. This statistic cannot be disputed. Anyone can look up the census reports and work it out with a pocket calculator. The census reports are to be found at https://www.ons.gov.uk/census. France does not record religious affiliation, but the trajectory would appear to be very similar.
94. "Europe's Growing Muslim Population," Pew Research Center, November 29, 2017, at https://www.pewforum.org/2017/11/29/europes-growing-muslim-population/.
95. For an acute analysis of the headscarves controversy in France, see John R. Bowen, *Why the French Don't Like Headscarves: Islam, the State and Public Space* (Princeton University Press, Princeton NJ, 2006). For the USA, see Darrell M. West, *Divided Politics, Divided Nation: Hyperpolitics in the Trump Era* (Brookings Institution Press, Washington DC 2019). For the roots of right-wing nationalism in the USA, see Anatol Lieven, *America Right or Wrong,* pages 81–155.
96. Collier, *Exodus,* page 7.

Chapter 3

1. Theodore Roosevelt, New Nationalism speech, Ossowatomie, Kansas, August 31, 1910, at https://teachingamericanhistory.org/library/document/new-nationalism-speech/.
2. David Goodhart, *The British Dream: Successes and Failures of Post-War Immigration* (Atlantic Books, London 2013), pages xxv, 7.
3. Paul Collier, *The Future of Capitalism: Facing the New Anxieties* (Allen Lane, London 2018), page 38; Dani Rodrik, *The Globalization Paradox: Democracy and the Future of the World Economy* (W. W. Norton, New York 2011), pages 207–232.
4. Eric Hobsbawm, *Nations and Nationalism* (Cambridge University Press, Cambridge 1990), page 38ff.; Avishai Margalit, "The Moral Psychology of Nationalism," in The *Morality of Nationalism,* ed. Robert McKim and Jeff McMahan (Oxford University Press, Oxford 1997).
5. Paul G. Harris, *What's Wrong with Climate Politics and How to Fix It* (Polity Press, Cambridge 2013), pages 33–63; Gregory White, *Climate Change and Migration* (Oxford University Press, New York 2011), page 6.
6. Rodrik, *Globalization Paradox,* page 208.
7. Perre Manent, *A World beyond Politics? A Defence of the Nation State,* translated by Marc LePain (Princeton University Press, Princeton NJ 2006), page 1.
8. For a contrary view, see Amartya Sen, *Identity and Violence: The Illusion of Destiny* (W. W. Norton, New York 2006).
9. David Miller, *On Nationality* (Clarendon Press, Oxford 1995), page 119; Paul Collier, *Exodus: How Migration Is Changing Our World* (Oxford University Press, Oxford 2013), page 263.

10. See John J. Mearsheimer, *The Great Delusion: Liberal Dreams and International Realities* (Yale University Press, New Haven CT 2018); Stephen M. Walt, *The Hell of Good Intentions: America's Foreign Policy Elite and the Decline of U.S. Primacy* (Macmillan, New York 2018); Andrew J. Bacevich, *Twilight of the American Century* (University of Notre Dame Press, Notre Dame IN 2018).
11. For a typical bipartisan establishment effort aimed at mobilizing liberalism behind US geopolitical agendas against Russia and China, see Bruce Jones and Torrey Taussig, eds., "Democracy and Disorder: The Struggle for Influence in the New Geopolitics," Brookings Institution, February 2019, at https://www.brookings.edu/wp-content/uploads/2019/02/FP_20190226_democracy_report_WEB.pdf. As usual in such productions, in 55 pages and 26,000 words, climate change is mentioned once.
12. See, for example, Andrew Sullivan, "The Establishment Will Never Say No to a War," *New York Magazine,* December 21, 2018, at https://medium.com/new-york-magazine/the-establishment-will-never-say-no-to-a-war-3af30630f23d.
13. Will Kymlicka, "The Sources of Nationalism," in McKim and McMahan, *The Morality of Nationalism*, page 57.
14. Liah Greenfeld, *Nationalism: Five Paths to Modernity* (Harvard University Press, Cambridge MA 2013), page 491.
15. Tom Nairn, *Faces of Nationalism* (Verso, London 1997), pages 65–67; Pierre Manent, *A World beyond Politics: A Defense of the Nation-State*, translated by Marc LePain (Princeton University Press, Princeton NJ 2006), pages 51–59.
16. Tom Nairn, "The Curse of Rurality," in *The State of the Nation: Ernest Gellner and the Theory of Nationalism*, ed. John A. Hall (Cambridge University Press, Cambridge 1998), page 108.
17. Delmer M. Brown, *Nationalism in Japan: An Introductory Historical Analysis* (University of California Press, Berkeley 1955), pages 91, 104.
18. See "The Imperial Rescript on Education" (1890), in *Sources of Japanese Tradition,* vol. 2, part 2, ed. William Theodore de Bary et al. (Columbia University Press, New York 2006), pages 108–110; Barrington Moore, *Social Origins of Dictatorship and Democracy: Lord and Peasant in the Making of the Modern World* (Penguin, London 1966), page 246ff.
19. Kevin M. Doak, *A History of Nationalism in Modern Japan: Placing the People* (Brill, Boston 2007), pages 36–45, 113–126.
20. James W. White, "State Building and Modernization: The Meiji Restoration," in *Crisis, Choice and Change: Historical Studies of Political Development*, ed. Gabriel A. Almond et al. (Little, Brown, Boston 1973), pages 502–503; de Bary, *Sources of Japanese Tradition*, pages 117–118.
21. Fukuzawa Yukichi, quoted in W. G. Beasley, *The Meiji Restoration* (Oxford University Press, Oxford 1973), page 377.

22. See Erica Benner, "Nationalism: Intellectual Origins," in *The History of Nationalism,* ed. John Breuilly (Oxford University Press, New York 2013), pages 36–51; James J. Sheehan, *German Liberalism in the Nineteenth Century* (Methuen, London 1982), pages 274–283.
23. Immanuel Wallerstein, *The Modern World-System IV: Centrist Liberalism Triumphant, 1789–1914* (University of California Press, Berkeley 2011), page 9.
24. Barrington Moore, *Social Origins of Dictatorship and Democracy: Lord and Peasant in the Making of the Modern World* (Penguin, London 1966) , page 493; Eugen Weber, *Peasants into Frenchmen: The Modernisation of Rural France, 1871–1914* (Chatto and Windus, London 1977); Heinz Ziegler, *Die Moderne Nation* (Tübingen 1931). For the relationship between the weakness of nationalism and state weakness in the Arab world, see Bassam Tibi, *Arab Nationalism: A Critical Enquiry,* translated by Marion Farouk-Sluglett (St. Martin's Press, New York 1990).
25. Francesco Trinchero, quoted in Nelson Moe, *The View from Vesuvius: Italian Culture and the Southern Question* (University of California Press, Berkeley 2002), page 145. For the authoritarian and military character of Italian liberalism after unification, see Moe, *The View from Vesuvius,* pages 126–183; Dennis Mack Smith, *The Making of Italy, 1796–1870* (Macmillan, London 1968), pages 371–394. For the Bronte revolt and its suppression, see Lucy Riall, *Under the Volcano: Revolution in a Sicilian Town* (Oxford University Press, Oxford 2013).
26. Francesco Trinchero, quoted in Nelson Moe, *The View from Vesuvius,* page 145. For the attitude of Russian liberals to the masses in the 1990s, see Anatol Lieven, Chechnya: *Tombstone of Russian Power* (Yale University Press, New Haven CT 1998), pages 153–155.
27. See Amy Chua, *Political Tribes: Group Instinct and the Fate of Nations* (Penguin Press, New York 2018), pages 161–173; Yasha Mounk, *The People versus Democracy* (Harvard University Press, Cambridge MA 2018), page 10ff.
28. Michael Kazin, "A Patriotic Left," Dissent, October 1, 2002.
29. Jill Lepore, This America: The Case for the Nation (Liveright, New York 2019), page 89.
30. For Kemalist ideology and Ataturk's reform program, see Sevket Pamuk, "Economic Change in Twentieth-Century Turkey," *Cambridge History of Turkey,* vol. 4 (Cambridge University Press, Cambridge, 2008), pages 266–300; Hugh Paulton, *Top Hat, Grey Wolf and Crescent: Turkish Nationalism and the Turkish Republic* (Hurst, London 1997), pages 92–129; Carter Vaughn Findley, *Turkey, Islam, Nationalism, and Modernity* (Yale University Press, New Haven CT 2010); William Hale, *The Political and Economic Development of Modern Turkey* (St. Martin's Press, New York 1981).

31. See Hasan Kayali, "The Struggle for Independence," and Andrew Mango, "Ataturk," in *The Cambridge History of Turkey*, ed. Resat Kasaba (Cambridge University Press, Cambridge 2006), pages 112–146, 147–174; Nicole and Hugh Pope, *Turkey Unveiled: Ataturk and After* (John Murray, London 1997), pages 50–69.
32. Anne Applebaum, "A Warning from Europe: The Worst Is Yet to Come," *The Atlantic*, September 13, 2018.
33. See Anatol Lieven, *The Baltic Revolution: Estonia, Latvia, Lithuania, and the Path to Independence* (Yale University Press, New Haven CT 1993), pages 316–318, 374–384.
34. See John Gray, *False Dawn: Delusions of Market Capitalism* (Granta Books, London 1998); Maurice Glasman, *Unnecessary Suffering: Managing Market Utopia* (Verso, London 1996).
35. See Stefan Auer, *Liberal Nationalism in Central Europe* (Routledge, London 2004).
36. Ivan Krastev, *After Europe* (University of Pennsylvania Press, Philadelphia 2017), pages 27–32.
37. Yael Tamir, *Why Nationalism* (Princeton University Press, Princeton NJ 2019), page 39. See also Krastev, *After Europe,* page 28. See also Bernard Yack, "The Liberal Democratic State," in T.V. Paul. G. John Ikenberry et al., *The Nation State in Question* (Princeton University Press, Princeton NJ 2003), page 36; Scruton, *Green Philosophy*, chap. 7, pages 90–107.
38. See, for example, Richard B. Howarth, "Intergenerational Justice," in *Oxford Handbook of Climate Change and Society*, ed. John S. Dryzek, Richard B. Norgaard, and David Schlosbert (Oxford University Press, New York 2011), pages 338–354; Nicholas Stern, *Why Are We Waiting?* (MIT Press, Cambridge MA 2015), chap. 6, pages 185–208.
39. Astra Taylor, "Time's Up for Capitalism but What Comes Next?," *The Nation*, May 6, 2019, at https://www.thenation.com/article/democracy-environment-astra-taylor/.
40. Aviel Roshwald, *The Endurance of Nationalism: Ancient Roots and Modern Dilemmas* (Cambridge University Press, New York 2006), page 51.
41. See the papal encyclical *Laudato Si* (2015) of Pope Francis, at https://laudatosi.com; for statements by previous popes on the environment and climate change, see https://catholicclimatemovement.global/statements-on-climate-change-from-the-popes/. For contemporary Catholic teaching on the environment, see http://www.usccb.org/issues-and-action/human-life-and-dignity/environment/environmental-justice-program/upload/Environmental-Primer.pdf. For wider religious thinking on the environment, see Thomas R. Dunlap, *Faith in Nature: Environmentalism as Religious Quest* (University of Washington Press, Seattle 2004); Dave Foreman, *Confessions of an Eco-Warrior* (Harmony Books, New York 1991); and Mike Hulme, *Why We Disagree about Climate Change* (Cambridge University Press, New York 2009), pages 142–177.

42. See William Nordhaus, *A Question of Balance: Weighting the Options on Global Warming Policies* (Yale University Press, New Haven CT 2008); Derek Parfit, "Energy Policy and the Further Future: The Identity Problem," in *Energy and the Future*, ed. Douglas MacLean and Peter G. Brown (Rowman and Littlefield, Totowa NJ 1983); Hulme, *Why We Disagree about Climate Change*, pages 118–126.
43. Stern, *Why Are We Waiting?*, pages 156–157.
44. David Wallace-Wells, *The Uninhabitable Earth: Life after Warming* (Tim Duggan Books, New York 2019), page 31.
45. Edmund Burke, *Reflections on the Revolution in France* (James Dodsley, London 1790), page 81. For the link to posterity as essential to the morale and "nerve" of a society, see Kenneth E. Boulding, "The Economics of the Coming Spaceship Earth," presented at the Sixth Resources for the Future Forum on Environmental Quality in a Growing Economy in Washington, DC, March 8, 1966, pages 10–11, at http://arachnid.biosci.utexas.edu/courses/THOC/Readings/Boulding_SpaceshipEarth.pdf.
46. Paul Ricoeur, *Figuring the Sacred: Religion, Narrative and the Imagination*, translated by David Pellauer, edited by Mark I. Wallace (Fortress Press, Minneapolis MN 1995), page 77.
47. Tom Nairn, *Faces of Nationalism*, page 4.
48. See Donald L. Horowitz, *Ethnic Groups in Conflict* (University of California Press, Berkeley 2000).
49. Virgil, *The Aeneid*, translated by W. F. Jackson Knight (Penguin Classics, London 1975), page 36.
50. Adam Tooze, "Rising Tides Will Sink Global Order," *Foreign Policy*, December 20, 2018.
51. Chandran Nair, *The Sustainable State* (Berrett-Koehler, Oakland CA 2018), pages 37–38.
52. Beatrice Pembroke and Ella Saltmarshe, "The Long Time," October 29, 2018, The Long Time Project, at https://medium.com/@thelongtimeinquiry; Roman Krznaric, "Why We Need to Reinvent Democracy for the Long Term," BBC, March 19, 2019, at http://www.bbc.com/future/story/20190318-can-we-reinvent-democracy-for-the-long-term.
53. For the Great Law of the Iroquois (which does not in fact mention any such thinking), see http://www.indigenouspeople.net/iroqcon.htm; Peter Wood, "Seventh Generation Sustainability—A New Myth?," National Association of Scholars, December 28, 2009, at https://www.nas.org/articles/Seventh_Generation_Sustainability_-_A_New_Myth. For Iroquois warfare, see Daniel K, Richter, "War and Culture: The Iroquois Experience," *William & Mary Quarterly*, vol. 40, no. 4, October 1983, pages 528–559. Many environmentalists may dream secretly of burning climate change deniers alive, but unlike the Iroquois with their prisoners, they would probably not actually put this into practice.

54. Prasenjit Duara, *The Crisis of Global Modernity: Asian Traditions and a Sustainable Future* (Cambridge University Press, New York 2014), page 282.
55. Duara, *The Crisis of Global Modernity*, page 29ff. See also Thomas Berry, *The Sacred Universe: Earth, Spirituality and Religion in the 21st Century* (Columbia University Press, New York 2009); Roger S. Gottlieb, *This Sacred Earth: Religion, Nature, Environment* (Routledge, Abingdon UK 2003), Hulme, *Why We Disagree about Climate Change*, pages 142–161. For contemporary attempts to bring together religious organizations and conservancy groups, see the Alliance for Religions and Conservation (ARC) at http://www.arcworld.org/.
56. Anthony D. Smith, *Nationalism and Modernism* (Routledge, London 1998), pages 181–187.
57. Garrett Mattingly, *The Defeat of the Spanish Armada* (Jonathan Cape, London 1970), page 336.
58. Miller, *On Nationality*, pages 36–37.
59. Cited in Richard Weight, *Patriots: National Identity in Britain, 1940–2000* (Pan, London 2003), page 183.
60. Theodore Roosevelt, *A Book-Lover's Holidays in the Open*, 1916 (reprinted CreateSpace Independent Publishing Platform 2015).
61. Harvey, *The Enigma of Capital and the Crises of Capitalism* (Profile Books, London 2011), page 184.
62. Avield Roshwald, *The Endurance of Nationalism* (Cambridge University Press, Cambridge 2011), page 67; Simon Schama, *Landscape and Memory* (Harper Perennial, London 2004); Robert Pogue Harrison, *Forests: The Shadow of Civilization* (University of Chicago Press, Chicago 1992); Eviatar Zerubavel, *Time Maps: Collective Memory and the Social Shape of the Past* (University of Chicago Press, Chicago 2003), page 41ff.
63. See Andrea Marks, "How the Mental Health Community Is Bracing for the Impact of Climate Change," *Rolling Stone*, May 16, 2019; Ciara O'Rourke, "Climate Change's Hidden Victim: Your Mental Health," *Medium*, January 24, 2019, at https://medium.com/s/2069/the-emotional-damage-done-by-climate-change-2f8f9ad59155; Wallace-Wells, *Uninhabitable Earth*, pages 136–138.
64. See Matthew Phelan, "The Menace of Eco-Fascism," *New York Review of Books*, October 22, 2018, at https://www.nybooks.com/daily/2018/10/22/the-menace-of-eco-fascism/; Sarah Manavis, "Eco-Fascism: The Ideology Marrying Environmentalism and White Supremacy Thriving Online," *New Statesman America*, September 21, 2018, at https://www.newstatesman.com/science-tech/social-media/2018/09/eco-fascism-ideology-marrying-environmentalism-and-white-supremacy.
65. For a passionate defense of *oikophilia* (love of home) against oikophobia among environmentalists, see Roger Scruton, *Green Philosophy: How to Think Seriously about the Planet* (Atlantic, London 2012), pages 104–107.

66. Burke, *Reflections*, page 81.
67. Mark Lilla, *The Once and Future Liberal: After Identity Politics* (Harper, New York 2017), page 15.
68. Yael Tamir, "Pro Patria Mori", in Robert McKim and Jeff McMahan, *The Morality of Nationalism* (Oxford University Press New York 1997), page 229. See also Isaiah Berlin, "The Bent Twig: A Note on Nationalism," *Foreign Affairs*, vol. 51, no. 1, October 1972, pages 11–30.
69. See Anatol Lieven, *America Right or Wrong: An Anatomy of American Nationalism* (2nd ed.) (Oxford University Press, New York 2012), pages 37–46; Will Kymlicka, "The Sources of Nationalism," page 58.
70. See Herbert Croly, *The Promise of American Life* (1909, reprinted with an introduction by Franklin Foer, Princeton University Press, Princeton NJ 2014); Richard Hofstadter, *The Progressive Movement, 1900–1915* (Spectrum, New York 1963); Sean Dennis Cashman, *America in the Gilded Age: From the Death of Lincoln to the Rise of Theodore Roosevelt* (New York University Press, New York 1984), pages 354–380.
71. John C. Harles, *Politics in the Lifeboat: Immigrants and the American Democratic Order* (Westview Press, Boulder CO 1993), page 100.
72. John Stuart Mill, *Considerations on Representative Government* (Floating Press, Auckland, New Zealand), page 344.
73. Christopher Caldwell, *Reflections on the Revolution in Europe* (Penguin, London 2010), pages 173–174, 195–199.
74. Stuart Hampshire, *Morality and Conflict* (Harvard University Press, Cambridge MA 1984), chap. 6.
75. Walter Russell Mead, *Special Providence: American Foreign Policy and How It Changed the World* (Knopf, New York 2001), pages 226–237; Lieven, *America Right or Wrong*, pages 37–46; Michael Lind, *The Next American Nation: The New Nationalism and the Fourth American Revolution* (Simon and Schuster, New York 1995), pages 270–274, 285–287.
76. Michael Walzer, in *Multiculturalism and the Politics of Recognition*, ed. Charles Taylor (Princeton University Press, Princeton NJ 1994), page 101; Kymlicka, *Politics in the Vernacular: Nationalism, Multiculturalism and Citizenship* (Oxford University Press, Oxford, 2001)
77. Robert Reich, *Common Good* (Penguin Random House, New York 2018), page 32. See also David Goodhart, *Road to Somewhere: The New Tribes Shaping British Politics* (Penguin, London 2017), pages 215–233.
78. For citizenship and duty, see David Selbourne, *The Principle of Duty* (Sinclair-Stevenson, London 1994), pages 222–243.
79. Stephen Holmes and Cass R. Sunstein, *The Cost of Rights: Why Liberty Depends on Taxes* (W. W. Norton, New York 1999), pages 15–16.
80. Yuval Levin, *The Fractured Republic: Renewing America's Social Contract in the Age of Individualism* (Basic Books, New York 2017), page 202.
81. For the decline of citizenship and its consequences, see Tamir, *Why Nationalism?*, pages 93–94; Mark Lilla, *Once and Future Liberal*, pages 131–133.

82. Claire Sutherland, *Nationalism in the Twenty-First Century: Challenges and Responses* (Palgrave Macmillan, London 2012), pages 18–19; Craig Calhoun, "Cosmopolitanism and Nationalism," *Nations and Nationalism,* vol. 13, no. 3, 2008, pages 427–448.
83. Pierre Manent, *A World beyond Politics? A Defense of the Nation State,* translated by Marc Le Pain (Princeton University Press, Princeton NJ 2006), page 205.
84. See Arlie Russsell Hochschildt, *Strangers in Their Own Land: Anger and Mourning on the American Right* (New Press, New York 2018), page 323; Reich, *Common Good,* pages 179–182.

Chapter 4

1. John Buchan, *A Lodge in the Wilderness* (William Blackwood and Sons, London 1906), page 219.
2. Earl Roberts, *Lord Robert's Message to the Nation* (John Murray, London 1912), page 45.
3. Cf. Bill McKibben, "A World at War: We Are under Attack from Climate Change—and Our Only Hope Is to Mobilize like We Did in World War II," *New Republic,* August 15, 2016, at https://newrepublic.com/article/135684/declare-war-climate-change-mobilize-wwii; Erika Bolstad, "World War Mentality Needed to Beat Climate Change," *Scientific American,* August 15, 2016, at https://www.scientificamerican.com/article/world-war-mentality-needed-to-beat-climate-change/; Paul Danish, "Fighting Climate Change Like World War II," *Boulder Weekly,* January 31, 2019; Michael Greiner, "A Concrete Proposal to Heal a Divided Nation," *Medium,* April 16, 2019, at https://medium.com/@GreinerOU/a-concrete-proposal-to-heal-a-divided-nation-b8c38fc31ae.
4. John A. Mathews, *Greening of Capitalism: How Asia Is Driving the Next Great Transformation* (Stanford University Press, Stanford CA 2015); Peter Newell, *Climate Capitalism: Global Warming and the Transformation of the Global Economy* (Cambridge University Press, Cambridge 2012).
5. Benjamin Kunkel, "How Much Is Too Much?," *London Review of Books,* vol. 33, no. 3, February 3, 2011.
6. For the socialist debate on de-growth versus the Green New Deal, see the following essays and exchanges in the *New Left Review* (*NLR*): Herman Daly and Benjamin Kunkel, "A Practical Programme for the Blue Planet," *NLR,* no. 109, January–February 2018, pages 81–106; Robert Pollin, "Against De-Growth: For a Green New Deal," *NLR,* no. 112, July–August 2018, pages 5–28; Mark Burton and Peter Somerville, "De-Growth: A Defence," *NLR,* no. 15, January–February 2019, pages 95–104; and Lola Seaton, "Green Questions," *NLR,* no. 15, January–February 2019, pages 105–130.
7. See "An Eco-Modernist Manifesto," signed by Mark Lynas and others, at http://www.ecomodernism.org/.

8. As set out, for example, in the papal encyclical on the environment, *Laudato Si*, of 2015.
9. Stephen Holmes and Cass Sunstein, *The Cost of Rights: Why Liberty Depends on Taxes* (W. W. Norton, New York 1999), page 231.
10. Karl Polanyi, *The Great Transformation: The Political and Economic Origins of Our Time* (Beacon Press, London 2002), page 3.
11. Michael B. Katz, *In the Shadow of the Poorhouse: A Social History of Welfare in America* (Basic Books, New York 1986), quoted in Theda Skocpol, *Protecting Soldiers and Mothers: The Origins of Social Policy in the United States* (reprinted Harvard University Press, Cambridge MA 1995), page 24.
12. Bernard Semmel, *Imperialism and Social Reform: English Social-Imperial Thought, 1895–1914* (Allen and Unwin, London 1960), pages 12–16, 209; Skocpol, *Protecting Soldiers and Mothers*. For these fears in Britain, see Anna Davin, "Imperialism and Motherhood," in *Tensions of Empire: Colonial Cultures in a Bourgeois World*, ed. Frederick Cooper and Ann Laura Stoler (University of California Press, Berkeley 1997), pages 87–151.
13. Gerhard A. Ritter, *Social Welfare in Germany and Britain* (Berg, Leamington Spa 1986); David Blackbourn and Geoff Eley, *The Peculiarities of German History* (Oxford University Press, New York 1984), pages 91–94; Gordon Craig, *Germany, 1866–1945* (Oxford University Press, Oxford 1981), pages 150–152; Albin Gladen, *Geschichte der Sozialpolitik in Deutschland* (Steiner Verlag, Stuttgart 1974), pages 1–85; Hans-Ulrich Wehler, *Bismarck und der Imperialismus* (1969, reprinted KiWi Bibliothek, Cologne 2017), page 459ff.
14. Quoted in Hans-Ulrich Wehler, *The German Empire 1871–1918* (Berg, Dover NH 1985), page 132.
15. See E. P. Hennock, "The Origins of British National Insurance and the German Precedent, 1880–1914," in *The Emergence of the Welfare State in Britain and Germany, 1850–1950*, ed. W. J. Mommsen (Routledge, London 1981), pages 84–106; George Dangerfield, *The Strange Death of Liberal England, 1910–1914* (reprint Transaction, London 2011), pages 219–221.
16. G. R. Searle, *The Quest for National Efficiency Study in British Politics and Political Thought, 1899–1914* (Blackhall, London 1999).
17. Quoted in Hennock, "The Origins of British National Insurance," page 88.
18. See Sidney Webb, *Twentieth-Century Politics: A Policy of National Efficiency* (Fabian Tract no. 108, London 1901); George Bernard Shaw, *Fabianism and Empire: A Fabian Manifesto* (Grant Richards, London 1900); Benjamin Kidd, *Individualism and After* (Herbert Spencer Lecture 1908, reprinted Cornell University Library, Ithaca NY 2009); Karl Pearson, *Social Problems: Their Treatment, Past, Present, and Future* (University of London Press, London 1912); Alfred Milner, *The Nation and the Empire* (Constable, London 1913); George Dangerfield, *The Strange Death of Liberal England, 1910–1914* (Serif reprint, London 2012); John Buchan, *A Lodge in the Wilderness* (1906, republished independently London 2018).

19. Robert James Scally, *The Origins of the Lloyd George Coalition: The Politics of Social Imperialism, 1900–1918* (Princeton University Press, Princeton NJ 2015), page 9.
20. See Richard Titmuss, "War and Social Policy," in Titmuss, *Essays on 'The Welfare State'* (reprinted Policy Press, London 2018), pages 44–53.
21. "Social Insurance and Allied Services," report by Sir William Beveridge to Parliament, November 1942, at https://sourcebooks.fordham.edu/mod/1942beveridge.asp; Sir William H. Beveridge, *The Pillars of Security; and Other Wartime Essays and Addresses* (reprinted Routledge, London 2015); Sir William H. Beveridge, *Full Employment in a Free Society* (reprinted Routledge, London 2015).
22. Semmel, *Imperialism and Social Reform*, pages 257–261.
23. Semmel, *Imperialism and Social Reform*, pages 257–261.
24. John Maynard Keynes, *How to Pay for the War* (1940), quoted in Larry Elliott et al., "A Green New Deal," page 28 (first report of the Green New Deal Group, UK, July 2008), at https://neweconomics.org/uploads/files/8f737ea195fe56db2f_xbm6ihwb1.pdf.
25. For classical historical studies of moral economies in specific populations, see E. P. Thompson, *Customs in Common* (Merlin Press, London 2010); James C. Scott, *The Moral Economy of the Peasant: Rebellion and Subsistence in Southeast Asia* (Yale University Press, New Haven CT 1977). For an extension of the concept to modern societies, see Norbert Goetz, "Moral Economy: Its Conceptual History and Analytical Prospects," *Journal of Global Ethics*, vol. 11, no. 2, 2015; Andrew Sayer, "Moral Economy and Political Economy," *Studies in Political Economy*, no. 61, 2000, pages 79–103.
26. See the opinion survey by the Levada Centre, March 2017, cited in Ian Bremmer, *Us vs. Them: The Failure of Globalism* (Penguin, New York 2018), page 138.
27. See Richard Hofstadter, *The Progressive Movement, 1900–1915* (Simon and Schuster, New York 1986); Michael Lind, *The Next American Nation: The New Nationalism and the Fourth American Revolution* (Simon and Schuster, New York 1995), pages 301–303; Michael McGerr, *A Fierce Discontent: Rise and Fall of the Progressive Movement, 1870–1920* (Oxford University Press, New York 2005).
28. Samuel Eliot Morison, Henry Steele Commager, and William E. Leuchtenberg, *The Growth of the American Republic*, vol. 2 (Oxford University Press, New York 1969), page 272.
29. Robert Reich, *Saving Capitalism: For the Many, Not the Few* (Vintage, New York 2016), pages 177–180.
30. Frank Norris, *The Octopus: A Story of California* (republished with an introduction by Kevin Starr, Penguin Books, New York 1986). For parallels to quasi-monopolistic practices (especially by Wall Street) today, see Reich, *Saving Capitalism*, pages 184–185.

31. Herbert Croly, *The Promise of American Life* (1909; republished with an introduction by Franklin Foer, Princeton University Press, Princeton NJ 2014) pages 28–29.
32. Reich, *Saving Capitalism*, pages 158–159.
33. President Theodore Roosevelt, "A Speech Delivered at the Dedication of the John Brown Memorial Park in Osawatomie," Kansas (known as the "New Nationalism speech"), August 31, 1910, at https://teachingamericanhistory.org/library/document/new-nationalism-speech/.
34. Roosevelt, "A Speech Delivered at the Dedication of the John Brown Memorial Park."
35. Barack Obama, "Remarks by the President at Osawatomie, Kansas, December 6, 2011," at https://obamawhitehouse.archives.gov/the-press-office/2011/12/06/remarks-president-economy-osawatomie-kansas.
36. Reich, *Saving Capitalism*, pages 175–176.
37. Holmes and Sunstein, *The Cost of Rights*; Paul Collier, *Exodus: How Migration Is Changing Our World* (Oxford University Press, New York 2015), pages 257–259.
38. Patrick J. Geary, *Before France and Germany: The Creation and Transformation of the Merovingian World* (Oxford University Press, New York 1988), page 29.
39. For the history of European states' struggle to raise revenue, see Charles Tilly, *Coercion, Capital and European States, AD 1990–1992* (Wiley, Hoboken NJ 2007). For contemporary states, this problem has been obvious for three decades, but no significant counter-measures have been achieved. See Christopher Hood, "The Tax State in the Information Age," in T.V. Paul. G. John Ikenberry et al., *The Nation State in Question* (Princeton University Press, Princeton N J 2003), pages 213–233.
40. See chapter 4, "The Masque of Democracy: Russia's Liberal Capitalist Revolution and the Collapse of State Power," in Anatol Lieven, *Chechnya: Tombstone of Russian Power* (Yale University Press, New Haven CT 1999), pages 147–185.
41. Franklin Foer, "Russian-Style Kleptocracy Is Infiltrating America," *The Atlantic*, March 2019.
42. Paul Collier, *The Future of Capitalism* (Allen Lane, London 2018), pages 193–194.
43. William G. Gale and Aaron Krupkin, "How Big Is the Problem of Tax Evasion," Brookings Institution, April 9, 2019, at https://www.brookings.edu/blog/up-front/2019/04/09/how-big-is-the-problem-of-tax-evasion/.
44. See Mathews, *Greening of Capitalism*.
45. Mathews, *Greening of Capitalism*, page 33.
46. Neta C. Crawford, "US Budgetary Costs of the Post-9/11 Wars through FY2019: $5.9 Trillion Spent and Obligated," Watson Institute, Brown University, Providence, RI, November 14, 2018.

47. As Rudyard Kipling said of the political influence of the great newspaper proprietors of his day (speaking through Prime Minister Stanley Baldwin), "What [they] are aiming at is power, and power without responsibility—the prerogative of the harlot throughout the ages," at http://www.thisdayinquotes.com/2011/03/power-without-responsibility.html.
48. For nuclear fusion, see W. J. Nuttall, *Nuclear Renaissance: Technologies and Policies for the Future of Nuclear Power* (Taylor and Francis, New York 2005), pages 239–304; Molly Lempriere, "Nuclear Fusion: Is Halfway Good Enough?," *Power Technology,* March 11, 2019, at https://www.power-technology.com/features/future-of-nuclear-fusion/; Hannah Devlin, "Nuclear Fusion on Brink of Being Realised, Say MIT Scientists," *The Guardian,* March 9, 2018, at https://www.theguardian.com/environment/2018/mar/09/nuclear-fusion-on-brink-of-being-realised-say-mit-scientists.
49. Robert Reich, *Common Good* (Penguin Random House, New York 2018) , pages 149–150; Bremmer, *Us vs. Them,* pages 143–145; Roman Lanis and Brett Govendir, "Three Strategies to Fight the Tax Avoidance Revealed in the Paradise Papers," *The Conversation,* 2019, at https://theconversation.com/three-strategies-to-fight-the-tax-avoidance-revealed-by-the-paradise-papers-87002.
50. See Linda Weiss, *America Inc? Innovation and Enterprise in the National Security State* (Cornell University Press, Ithaca NY 2014).
51. Nicholas Stern, *Why Are We Waiting?: The Logic, Urgency and Promise of Tackling Climate Change* (MIT Press, Cambridge MA 2015), page 62.
52. Stern, *Why Are We Waiting?*, page 33. For another optimistic view of the profits and jobs to be created in the move to low-carbon economies, see "The New Climate Economy," The Global Commission on the Economy and Climate (GCEC), 2018, at https://newclimateeconomy.report/2018/.
53. Kingsmill Bond, "2020 Vision," *Carbon Tracker*, September 10, 2018, at https://www.carbontracker.org/reports/2020-vision-why-you-should-see-the-fossil-fuel-peak-coming/,
54. Mathews, *Greening of Capitalism*, pages 222–223.
55. See Deirdre Nansen McCloskey's home page at https://www.deirdremccloskey.com/.
56. David Wallace-Wells, *The Uninhabitable Earth: A Story of the Future* (Allen Lane, London 2019), page 237.
57. "India on Track to Achieve 175 GW of Renewable Energy by 2022," *Times of India*, October 15, 2019, at https://timesofindia.indiatimes.com/india/india-on-track-to-achieve-175-gw-of-renewable-energy-by-2022-union-minister/articleshow/71598208.cms.
58. Stern, *Why Are We Waiting?*, pages 38, 94–95.
59. "Global Trends in Renewable Energy Report," *Green Growth Knowledge Platform*, April 2018, at https://about.bnef.com/blog/

clean-energy-investment-exceeded-300-billion-2018/; BloombergNEF, "Clean Energy Investment Exceeded $300 Billion Again in 2018," January 16, 2019, at https://about.bnef.com/blog/clean-energy-investment-exceeded-300-billion-2018/; Stern, *Why Are We Waiting?*, pages 76–77.

60. Mathews, *Greening Capitalism*, pages 42–48.
61. For somewhat rosy-eyed views of the wider advantages of the Chinese system, see Hu Angang, *China in 2020: A New Kind of Superpower* (introduction by Cheng Li, Brookings Institution Press, Washington DC 2011), pages 121–162; Daniel A. Bell, *The China Model: Political Meritocracy and the Limits of Democracy* (Princeton University Press, Princeton NJ 2015). See also Martin Jacques, *When China Rules the World: The Rise of the Middle Kingdom and the End of the Western World* (Allen Lane, London 2009).
62. Collier, *Future of Capitalism*, pages 69–85.
63. Richard Roberts, "A Six-Point Plan for Building a New Carbon Economy," *Medium*, September 14, 2018, at https://medium.com/fast-company/a-6-point-plan-for-building-a-new-carbon-economy-e5e122d24612.
64. Pallab Ghosh, "Climate Change: Scientists Test Radical Ways to Fix Earth's Climate," BBC Science and Environment, May 10, 2019, at https://www.bbc.com/news/science-environment-48069663.
65. Andy Skuce, "Why Carbon Capture Is No Panacea," *Bulletin of Atomic Scientists*, October 4, 2016, at https://thebulletin.org/2016/10/wed-have-to-finish-one-new-facility-every-working-day-for-the-next-70-years-why-carbon-capture-is-no-panacea/; Stern, *Why Are We Waiting?*, pages 63–64; Sang Jing, "Energy Efficiency: China's 'First Fuel' Lighting Up a Sustainable Pathway," *International Partnership for Energy Efficiency Cooperation* (IPEEC), February 11, 2019, at https://ipeec.org/bulletin/98-energy-efficiency-chinas-first-fuel-lighting-up-a-sustainable-pathway.html; Stern, *Why Are We Waiting?*, page 45.
66. See Jeff Goodell, "Can Civilization Survive What's Coming?," *Rolling Stone*, October 9, 2018, at https://medium.com/rollingstone/can-civilization-survive-whats-coming-7155fb2ee1c7.
67. Seth M. Siegel, *Let There Be Water: Israel's Solution for a Water-Starved World* (St. Martin's Press, New York 2017); Scott Michael Moore, "Israel: How Meeting Water Challenges Spurred a Dynamic Export Industry," World Bank blog, October 27, 2017, at http://blogs.worldbank.org/water/israel-how-meeting-water-challenges-spurred-dynamic-export-industry.
68. David Hazony, "How Israel Is Solving the Global Water Crisis," *The Tower*, issue 31, October 2015, at http://www.thetower.org/article/how-israel-is-solving-the-global-water-crisis/.
69. Stern, *Why Are We Waiting?*, pages 105–106.
70. Collier, *Future of Capitalism*, page 49.

Chapter 5

1. Allan Bloom, *The Closing of the American Mind* (Simon and Schuster, New York 1987), page 27.
2. Russell Kirk, "Common Reader for Everyday Ecologists," *New Orleans Times-Picayune*, September 20, 1971.
3. "Plan d'action: Sauver l'Europe pour sauver le climat," program for the European elections, May 2019, at https://www.pourleclimat.eu/le-plan-daction.
4. "Nous avons droit au bonheur. Nous avons le droit de vouloir le meilleur pour celles et ceux que nous aimons," Profession de Foi, Europe Ecologie-Les Verts, European elections, May 2019, at https://programme-candidats.interieur.gouv.fr/data-pdf-propagandes.
5. "Construisons-la aujourd'hui autour de l'objectif de sortir du carbone et du nucléaire et d'investir massivement dans la transition écologique. . . . Grâce à une politique d'asile réformée et unifiée, construisons une Europe capable d'assumer la question de l'accueil des hommes et des femmes que le fracas du monde a jeté-e-s sur les routes au péril de leur vie, une Europe capable de continuer à assurer la paix sur le continent," Plan d'Action: Sauver l'Europe pour Sauver le Climat, European elections, May 2019, at https://www.pourleclimat.eu/le-plan-daction.
6. For nuclear power and climate change, see W. J. Nuttall, *Nuclear Renaissance: Technologies and Policies for the Future of Nuclear Power* (Taylor and Francis, New York 2005), pages 54–60.
7. For the program of the British Green Party, see "Our Political Programme," 2019, at https://www.greenparty.org.uk/assets/images/national-site/political-programme-web-v1.3.pdf.
8. David Wallace-Wells, *The Uninhabitable Earth: A Story of the Future* (Allen Lane, London 2019), pages 181–185; James Lovelock, "Nuclear Power for the 21st Century," May 21, 2005, at http://www.jameslovelock.org/nuclear-energy-for-the-21st-century/.
9. "The Chernobyl Accident: UNSCEAR's assessment of the radiation effects", UNSCEAR 2012, at https://www.unscear.org/unscear/en/chernobyl.html.
10. Maureen C. Hatch et al., "Cancer Near the Three Mile Island Nuclear Plant: Radiation Emissions", *American Journal of Epidemiology*, September 1990, Vol. 132, Issue 3, September 1990, pages 397–412, at https://doi.org/10.1093/oxfordjournals.aje.a115673; Steve Wing et al., "A reevaluation of cancer incidence near the Three Mile Island nuclear plant: the collision of evidence and assumptions", *Environmental Health Perspectives*, January 1997, Vol. 105, No. 1, pages 52–57, at https://www.ncbi.nlm.nih.gov/pmc/articles/PMC1469835/.
11. Wallace-Wells, *Uninhabitable Earth*, page 182.
12. "Comparing Nuclear Accident Risks with Those from Other Energy Sources," Organisation for Economic Co-operation and Development

(OECD), August 31, 2010, at http://www.oecd.org/publications/comparing-nuclear-accident-risks-with-those-from-other-energy-sources-9789264097995-en.htm; Nicholas Stern, *Why Are We Waiting?: The Logic, Urgency and Promise of Tackling Climate Change* (MIT Press, Cambridge MA 2015), pages 123–125.

13. "Chernobyl: The True Scale of the Accident," United Nations Information Service, September 6, 2005, at https://www.un.org/press/en/2005/dev2539.doc.htm; "Levels and Effects of Radiation Exposure Due to the Nuclear Accident after the 2011 Great East Japan Earthquake and Tsunami," United Nations Scientific Committee on the Effects of Atomic Radiation (UNSCEAR), 2013, at http://www.unscear.org/docs/reports/2013/13-85418_Report_2013_Annex_A.pdf; "FAQs: Fukushima Five Years On," World Health Organization (WHO), 2017, at https://www.who.int/ionizing_radiation/ae/fukushima/faqs-fukushima/en/.
14. German Green Party chair Claudia Roth, talk at Georgetown University in Qatar, April 10, 2019.
15. Wily Blackmore, "Germany Has a Radioactive Wild Boar Problem—and It's Chernobyl's Fault," *Takepart*, September 2, 2014, at http://www.takepart.com/feature/2014/09/02/radioactive-wild-boars.
16. James Lovelock, "Nuclear Power Is the Only Green Solution," *The Independent*, May 24, 2004. See also Joshua S. Goldstein and Staffan Qvist, *A Bright Future: How Some Countries Have Solved Climate Change and Others Can Follow* (Public Affairs, New York 2019).
17. For the German Alliance 90/Green Party platform, see https://www.gruene-bundestag.de/service-navigation/english.html.
18. For the statement on climate change in the Alternative for Germany (AfD) program, see *Programm der Alternative fuer Deutschland fuer die Wahl zum 9. Europaeischen Parlament 2019,* page 79, at https://www.afd.de/wp-content/uploads/sites/111/2019/02/AfD_Europawahlprogramm_A5-hoch_RZ.pdf.
19. For the relatively (but only relatively) more sensible approach of the French Front Nationale, see "French Election 2017: Where the Candidates Stand on Energy and Climate Change," *Carbon Brief*, March 7, 2017, at https://www.carbonbrief.org/french-election-2017-where-candidates-stand-energy-climate-change. For a ranking of the European parties on their commitment to climate change action, see "Defenders, Delayers, Dinosaurs: Ranking of EU Political Groups and National Parties on Climate Change," Climate Action Network Europe, April 2019, at http://www.caneurope.org/docman/climate-energy-targets/3476-defenders-delayers-dinosaurs-ranking-of-eu-political-groups-and-national-parties-on-climate-change/file. I would, however, disagree with this ranking, since if you factor in the support of the Green parties for the abolition of nuclear energy, their position drops from the top of the chart to somewhere in the middle.

20. Naomi Klein, *This Changes Everything: Capitalism versus the Climate* (Simon and Schuster, New York 2014).
21. See Marc Jaccard's review of Naomi Klein's work, "I Wish This Changed Everything: Is a Radical Economic Overhaul Our Best Chance to Save the Climate?," *Literary Review of Canada*, November 2014, at https://reviewcanada.ca/magazine/2014/11/i-wish-this-changed-everything/.
22. For the text of the resolution, see House Resolution 109, 116th Congress, February 7, 2019, "Recognizing the Duty of the Federal Government to Create a Green New Deal," at https://www.congress.gov/bill/116th-congress/house-resolution/109/text.
23. Natasha Geiling, "The Democratic Party Has a Climate Change Problem," Sierra Club, August 22, 2018, at https://www.sierraclub.org/sierra/democratic-party-has-climate-change-problem.
24. See Lisa Hagen and Timothy Cama, "DNC Reverses Ban on Fossil Fuel Donations," *The Hill*, August 10, 2018, at https://thehill.com/policy/energy-environment/401356-dnc-passes-resolution-on-fossil-fuel-donations; Haley Britzky, "DNC Now Accepting Donations from Fossil Fuel Workers and PACs," *Axios*, August 11, 2018, at https://www.axios.com/dnc-now-accepting-donations-from-fossil-fuel-workers-and-pac.
25. For Schwarzenegger's statements and policies on climate change as governor, see http://www.schwarzeneggerinstitute.com/institute-in-action/energy-and-the-environment. For a particularly pugnacious statement on climate change by the former Governator, see Arnold Schwarzenegger, "I don't give a XXXX if we agree on climate change," https://www.facebook.com/notes/arnold-schwarzenegger/i-dont-give-a-if-we-agree-about-climate-change/10153855713574658/; interview with Schwarzenegger in James Cameron's *Story of Science Fiction* (Insight Editions, San Rafael CA 2018), page 210. For a socialist argument for automation, see Casey Williams, "The Socialist Case for Automating Our Jobs Away," May 9, 2019, at https://onezero.medium.com/the-socialist-case-for-automating-our-jobs-away-4a12780309d2.
26. See the group's website at https://www.greennewdealgroup.org/. See also David Mackay, *Alternative Energy: Without the Hot Air* (Green Books, Cambridge UK 2009).
27. *Briefing Book for a New Administration: Recommended Policies and Practices for Addressing the Security Risks of a Changing Climate,* Admiral Frank Bowman et al., Climate and Security Advisory Group, September 14, 2016, at https://climateandsecurity.files.wordpress.com/2016/09/climate-and-security-advisory-group_briefing-book-for-a-new-administration_2016_11.pdf.
28. See "A National Strategic Narrative: A Vision for America in the Age of Uncertainty," by "Mr Y," originally published in *Foreign Policy*, April 13, 2011, at https://www.wilsoncenter.org/sites/default/files/A%20National%20Strategic%20Narrative.pdf.

29. See Oliver Milman, "The Young Republicans Breaking with Their Party over Climate Change," *The Guardian,* April 12, 2019, at https://www.theguardian.com/environment/2019/apr/11/we-can-talk-about-this-millennial-republicans-take-a-methodical-approach-to-climate-change.
30. See Robinson Meyer, "The Green New Deal Has Already Won," *The Atlantic*, June 5, 2019, at https://www.theatlantic.com/science/archive/2019/06/bidens-climate-plan-mini-green-new-deal/591046/; Justin Worland, "The Senate Will Reject the Green New Deal, but It Is Already Changing the Debate on Climate Change," *Time*, March 27, 2019, at http://time.com/5558370/green-new-deal-senate-vote/; Paul Krugman, "Purity versus Pragmatism, Environment versus Health," *New York Times,* April 11, 2019, at https://www.counterpunch.org/2019/02/15/what-a-green-new-deal-should-look-like-filling-in-the-details/ . For intelligent business responses to the Green New Deal idea, see Bloomberg's 2018 climate change analysis: https://data.bloomberglp.com/company/sites/48/2019/04/ScenarioAnalysis.pdf. For Bloomberg's take on the Green New Deal, see Noah Smith, "How to Design a Green New Deal That Isn't over the Top," *Bloomberg Opinion*, February 12, 2019, at https://www.bloomberg.com/opinion/articles/2019-02-12/an-alternative-to-alexandria-ocasio-cortez-s-green-new-deal.
31. For the platforms of the leading candidates as of June 2019, see, for Joe Biden, https://joebiden.com/Climate/; for Elizabeth Warren, https://elizabethwarren.com/issues; for Bernie Sanders, https://berniesanders.com/; for Beto O'Rourke, https://betoorourke.com/.
32. See the interview of Professor Robert Pollin with C. J. Polychroniou, *Global Policy,* February 8, 2019, at https://www.globalpolicyjournal.com/blog/08/02/2019/heres-what-green-new-deal-looks-practice.
33. Valerie Volcovici, "Labor Unions Fear Democrats' Green New Deal Poses Job Threats," Reuters, February 12, 2019, at https://www.reuters.com/article/us-usa-greennewdeal-coal/labor-unions-fear-democrats-green-new-deal-poses-job-threat-idUSKCN1Q11D2; "AFL-CIO Criticizes Green New Deal, Calling It not Achievable or Realistic," *Washington Post,* March 12, 2019; Rachel M. Cohen, "Labor Unions Are Skeptical of the Green New Deal, and They Want Activists to Hear Them Out," *The Intercept,* February 28, 2019, at https://theintercept.com/2019/02/28/green-new-deal-labor-unions/.
34. See Elizabeth Warren, "Our Military Can Help Lead the Fight against Climate Change," May 15, 2019, at https://medium.com/@teamwarren/our-military-can-help-lead-the-fight-in-combating-climate-change-2955003555a3; Timothy Gardner, "US Military Marches Forward on Green Energy, Despite Trump," Reuters, March 1, 2017, at https://www.reuters.com/article/us-usa-military-green-energy-insight/u-s-military-marches-forward-on-green-energy-despite-trump-idUSKBN1683BL; Saltanat Berdikeeva, "The US Military: Winning the Renewable War,"

Energy Digital, September 13, 2017, at https://www.energydigital.com/renewable-energy/us-military-winning-renewable-war.

35. Thomas L. Friedman and Michael Mandelbaum, *This Used to Be Us: How America Fell behind in the World We Created and How We Can Come Back* (Picador, New York 2012).

36. See Lawrence M. Krauss, "The World Needs a Terrestrial Sputnik Moment," *The Atlantic,* October 4, 2017, at https://www.theatlantic.com/science/archive/2017/10/the-world-needs-a-terrestrial-sputnik-moment/541989/; Manu Saadia, "Is the US Facing Another New Sputnik Moment?," *New Yorker,* October 4, 2017. In 2014, EU commissioner for research, science, and innovation Carlos Moedas called for an EU "Sputnik Moment" along these lines: Henry Goodwin, "Europe Facing a Sputnik Moment, Says Commissioner Moedas," *European University Institute,* October 6, 2017, at https://times.eui.eu/europe-facing-sputnik-moment-says-eu-commissioner-moedas.html; US energy secretary Steven Chu called for such a clean energy Sputnik Moment in competition with China back in 2010—but the Obama administration did far too little and the Trump administration has done nothing at all: Suzanne Goldenberg, "US Energy Secretary Warns of 'Sputnik Moment' in Green Technology Race," *The Guardian*, November 29, 2010, at https://www.theguardian.com/world/2010/nov/29/us-green-technology-energy-investment; Shaun Tandon, "US: China Rise a 'Sputnik Moment' for Clean Energy," Physics.org, November 30, 2010, at https://phys.org/news/2010-11-china-sputnik-moment-energy.html. See also Jonathan Schell, "The Sputnik Moment That Wasn't," *The Nation*, January 20, 2011.

37. Mark Lilla, *The Once and Future Liberal: After Identity Politics* (Harper, New York 2017), page 29.

38. Herbert Croly, *The Promise of American Life* (1909, reprinted with an introduction by Franklin Foer (Princeton University Press, Princeton NJ 2014), page 28 and passim.

39. Quoted in George Evans-Jones, "An Overarching, Simple Message That the Democrats Can Win On; Let's Make America Feel Great Again," *Medium*, March 5, 2019, at https://medium.com/@georgeej111/an-overarching-simple-message-that-the-democrats-can-win-on-a11480016f55.

40. John Rynn, "What a Green New Deal Should Look Like: Filling in the Details," *Counterpunch,* February 15, 2019, at https://www.counterpunch.org/2019/02/15/what-a-green-new-deal-should-look-like-filling-in-the-details/.

41. Quoted in Michael Beschloss, "What Took Them So Long?," *Newsweek*, November 20, 1994.

42. See Thomas Frank, *Listen, Liberal: Or, Whatever Happened to the Party of the People* (Harper, New York 2017), pages 52–59.

43. Garrett Hardin, "Tragedy of the Commons," *Science*, no. 162, 1968, pages 1243–1248.

44. See, for example, the "Intersectionality Calculator," on the web at https://intersectionalityscore.com/learn. According to this site, the richest, most able-bodied, and most highly educated black woman has an "intersectionality score" of 57, meaning "80 percent of others are more privileged than you." The richest, most highly educated white woman scores 38, meaning "70 percent of others are more privileged than you." The poorest, least educated, and most disabled white male scores 21, meaning "you are more privileged than 57 percent of others."
45. David Rosenberg, "Ocasio-Cortez Green New Deal Is Surest Way to Lose War on Climate Change," *Haaretz*, May 15, 2019, at https://www.haaretz.com/us-news/.premium-ocasio-cortez-s-green-new-deal-is-surest-way-to-lose-war-on-climate-change-1.6935504. For the history of the Democratic Party's abandonment of the working classes and removal of their representatives from the party leadership, see Thomas Frank, *Listen, Liberal!*, page 44ff.
46. "Income and Wealth in the United States: An Overview of Recent Data," Peter G. Peterson Foundation, September 13, 2018, at https://www.pgpf.org/blog/2018/09/income-and-wealth-in-the-united-states-an-overview-of-data.
47. Hawkins, S., et al., *Hidden Tribes: A Study of America's Polarized Landscape. More in Common*, at https://hiddentribes.us/pdf/hidden_tribes_report.pdf. See also Yascha Mounk, "Americans Strongly Dislike PC Culture," *The Atlantic*, October 10, 2018, at https://www.theatlantic.com/ideas/archive/2018/10/large-majorities-dislike-political-correctness/572581/. For a liberal protest against liberal cultural prescriptiveness, see Henry Wismayer, "Liberals, Please Chill Out," *Medium*, June 16, 2018, at https://medium.com/s/jeremiad/liberals-please-chill-out-7f7309e4d364.
48. Frank, *Listen, Liberal!*, pages 59–126: "To judge by what he actually accomplished, Bill Clinton was not the lesser of two evils, as people on the Left always say about Democrats at election time. He was the greater of the two. What he did as president was beyond the reach of even the most diabolical Republican. Only smiling Bill Clinton, well-known friend of working families, could commit such betrayals" (page 122).
49. Robert Reich, *Saving Capitalism: For the Many, Not the Few* (Vintage, New York 2016), pages 187–189.
50. Yasha Mounk, *The People Versus Democracy* (Harvard University Press, Cambridge MA 2018), page 17.
51. Jonathan Haidt, *The Righteous Mind: Why Good People Are Divided by Politics and Religion* (Pantheon Books, New York 2012). As Haidt writes, this is hardly an illiberal perception. Apart from it being what we know of ourselves, if we have a modicum of self-awareness, it goes back at the very least to David Hume and *An Enquiry Concerning Human Understanding* (1748, republished by Oxford University Press, Oxford 2008, with an introduction by Peter Millican).

52. Ursula K. Le Guin, *The Wind's Twelve Quarters* (Bantam, London 1976), pages 166, 251.
53. Jeremiah 8:12–13.
54. For a personal account of this, see J. D. Vance, *Hillbilly Elegy: A Memoir of a Family and Culture in Crisis* (Harper, New York 2018). For the history of the white "underclass" and elite attitudes toward it, see Nancy Isenberg, *White Trash: The 400-Year History of Class in America* (Atlantic Books, London 2017). For a book by a famous (or notorious) conservative pointing out the similarity of social and familial disintegration among poor whites and poor blacks, see Charles Murray, *Coming Apart: The State of White America, 1960–2010* (Crown Forum, New York 2013).
55. Tucker Carlson, "Mitt Romney Supports the Status Quo. But for Everybody Else, It's Infuriating," Fox News, January 3, 2019; see also "A Future to Believe In?," Isaac Wilks, *The Politic,* April 10, 2019, at https://thepolitic.org/afuture-to-believe-in/.
56. See Thomas Frank, *Listen, Liberal!*, page 39.
57. Yuval Levin, *The Fractured Republic: Renewing America's Social Contract in the Age of Individualism* (Basic Books, New York 2017) , pages 223–224.
58. See Kevin D. Williamson, "The Father-Fuehrer," *National Review*, March 28, 2016, at https://www.nationalreview.com/magazine/2016/03/28/father-f-hrer/.
59. Barbara Ehrenreich, "Dead, White and Blue: The Great Die-Off of America's Blue-Collar Whites," December 1, 2015, at https://www.guernicamag.com/barbara-ehrenreich-dead-white-and-blue/. For an example of such liberal contempt for poor whites, see W. Island, "Democrats Should Stop Trying to Win Over White Voters," *Medium*, November 25, 2018, at https://medium.com/we-know-what-we-know/democrats-should-stop-trying-to-win-over-rural-voters-340919153827.
60. Cited in Arlie Russsell Hochschildt, *Strangers In Their Own Land: Anger and Mourning on the American Right* (New Press, New York 2018), page 323.
61. Martin Luther King Jr., *Where Do We Go from Here? Chaos of Community*, introduction by Vincent Harding (reprinted Beacon Press, Boston MA 2010), pages 51–52.
62. See, for example, Umair Haque, "The Anglo World Is Collapsing: How the Dunces of Modern History Ended Up Being Us," *Medium,* March 27, 2019, at https://eand.co/why-the-anglo-world-is-collapsing-2223733ba0f9; Sophie Lewis, *Full Surrogacy Now: Feminism against Family* (Verso, New York 2019).
63. See Elizabeth Warren, "A Plan for Economic Patriotism," at https://medium.com/@teamwarren/a-plan-for-economic-patriotism-13b879f4cfc7. See also Robert Kuttner, "Warren's Astonishing Plan: Economic Patriotism," *American Prospect*, June 4, 2019, at https://prospect.org/article/warrens-astonishing-plan-economic-patriotism.

64. President Theodore Roosevelt, seventh message to Congress, 1907.
65. President Dwight D. Eisenhower, "Chance for Peace" speech (to American Society of Newspaper Editors), April 16, 1953, at https://www.eisenhowerlibrary.gov/all_about_ike/speeches/chance_for_peace.pdf.
66. The Democrats could also quote the passage on conservation from Eisenhower's farewell address; and from more recent speeches on the danger of climate change of Republicans like Senator John McCain.
67. Croly, *Promise of American Life*, pages 323–324.
68. David Shearman and Joseph Wayne Smith, *The Climate Change Challenge and the Failure of Democracy* (Praeger, Santa Barbara CA, 2007).

Conclusions

1. For the classic statement of the distinction between an ethic of conviction, or attitude possible for a private individual (Gesinnungsethik) and the ethic of responsibility necessary for a public official, see Max Weber, "Politics as a Vocation," 1919, at https://archive.org/details/weber_max_1864_1920_politics_as_a_vocation. See also Anatol Lieven and John Hulsman, *Ethical Realism: A Vision for America's Role in the World* (Pantheon Books, New York 2006), pages 77–79.
2. Anatol Lieven, *America Right or Wrong: An Anatomy of American Nationalism*, 2nd ed. (Oxford University Press, New York 2012), pages 37–51.
3. See Stephen M. Walt, *The Hell of Good Intentions: America's Foreign Policy Elite and the Decline of U.S. Primacy* (Farrar, Straus and Giroux, New York 2018), pages 21–90; Mearsheimer, *Great Delusion*, pages 1–44.
4. "One has the uneasy feeling that America as both a powerful nation and a 'virtuous' one is involved in ironic perils which compound the experiences of Babylon and Israel." Reinhold Niebuhr, *The Irony of American History* (1952, republished Scribner, New York 1985), page 160. See also Hans Morgenthau and Kenneth W. Thompson, *Politics Among Nations: The Struggle for Power and Peace* (MCGraw-Hill, New York, 2005), page 10. We can't say we weren't warned.
5. David Goodhart, *The British Dream: Successes and Failures of Postwar Immigration* (Atlantic Books, London 2013), page 25, and *The Road to Somewhere: The Populist Revolt and the Future of Politics* (Hurst, London 2017), page 15.
6. See Walt, *Hell of Good Intentions*, pages 181–216.
7. Stephen Holmes and Cass R. Sunstein, *The Cost of Rights: Why Liberty Depends on Taxes* (W. W. Norton, New York 1999), page 23–24.
8. Johann Wolfgang von Goethe, quoted in Hans J. Morgenthau, *Scientific Man versus Power Politics* (University of Chicago Press, Chicago 1946), page 218.
9. A. J. H. Murray, "The Moral Politics of Hans Morgenthau," *Review of Politics*, vol. 58, no. 1, Winter 1996, pages 81–108.

10. Quoted in Ivan Krastev, *After Europe* (University of Pennsylvania Press, Philadelphia 2017), page 7.
11. See Lieven, *America Right or Wrong*, pages 65–66.
12. For the ways in which people of the future may judge our contemporary literature and art in the context of climate change, see Amitav Ghosh, *The Great Derangement: Climate Change and the Unthinkable* (University of Chicago Press, Chicago 2016), page 11.
13. Hans Morgenthau, quoted in Robert C. Good, "The National Interest and Political Realism: Niebuhr's 'Debate' with Morgenthau and Kennan, *The Journal of Politics*, vol. 22, no. 4 (1960), page 612.
14. See Hugh White, *The China Choice: Why We Should Share Power* (Oxford University Press, Oxford 2012), pages 135–146.
15. Kenneth E. Boulding, "The Economics of the Coming Spaceship Earth," in *Environmental Quality in a Growing Economy*, ed. Henry Jarrett (Johns Hopkins University Press, Baltimore MD 1968), at http://arachnid.biosci.utexas.edu/courses/THOC/Readings/Boulding_SpaceshipEarth.pdf; Garrett Hardin, "Lifeboat Ethics," at http://web.ntpu.edu.tw/~language/course/research/lifeboat.pdf; "The Tragedy of the Commons," *Science*, vol. 162, no. 3859, December 13, 1968, pages 1243–1248, at http://arachnid.biosci.utexas.edu/courses/THOC/Readings/Boulding_SpaceshipEarth.pdfhttps://sites01.lsu.edu/faculty/kharms/wpcontent/uploads/sites/23/2017/04/HardinG_1968_Science.pdf.
16. As in Ursula Le Guin, "Paradises Lost," in *The Birthday of the World* (Harper Collins, New York 2002), pages 249–362. And even then, being humans, most of them end up by developing a fanatical religion under theocratic leadership, abandoning their assigned mission of exploration and settlement, and vanishing from the ken of the rest of humanity.

INDEX

Page followed by n indicate Endnotes.

For the benefit of digital users, indexed terms that span two pages (e.g., 52–53) may, on occasion, appear on only one of those pages.